Surface-Based Remote Sensing of the Atmospheric Boundary Layer

ATMOSPHERIC AND OCEANOGRAPHIC SCIENCES LIBRARY

Volume 40

For further volumes:
http://www.springer.com/series/5669

Stefan Emeis

Surface-Based Remote Sensing of the Atmospheric Boundary Layer

Springer

Dr. Stefan Emeis
Karlsruhe Institute of Technology
Institute for Meteorology and Climate
Research – Atmospheric Environmental
Research (IMK-IFU)
Garmisch-Partenkirchen
Germany
stefan.emeis@kit.edu

ISSN 1383-8601
ISBN 978-90-481-9339-4 e-ISBN 978-90-481-9340-0
DOI 10.1007/978-90-481-9340-0
Springer Dordrecht Heidelberg London New York

Library of Congress Control Number: 2010933507

Cover illustration: Wave@2009 JupiterImages Corporation

Printed on acid-free paper

Springer is part of Springer Science–Business Media (www.springer.com)

Preface

There exist several good books on the atmospheric boundary layer (ABL) from a theoretical or fluid dynamics point of view starting with Stull's book from 1988. A very recent example (although in the German language) is Kraus' book *Grundlagen der Grenzschicht-Meteorologie*, which was just published by Springer in 2008. Books which concentrate on experimental ABL research are less frequent.

Most experimental ABL research has been made so far by surface-based in situ experimentation (tower measurements up to a few hundred metres, surface energy balance measurements, short aircraft experiments, short experiments with tethered balloons, constant-level balloons, evaluation of radiosonde data). A good book on experimental surface-based in situ ABL research is the book *Micrometeorology* by Foken (2008), which concentrates on surface flux measurements.

Although the surface fluxes are one of the main driving factors for the daily variation of the ABL, an ABL description is only complete if its complete vertical structure is analysed and determined. Satellite information is available covering large areas, but it has only limited temporal resolution and lacks sufficient vertical resolution. Therefore, surface-based remote sensing is probably the only possibility to enlarge the database for ABL studies, as it offers nearly continuous and vertically highly resolved information for specific sites of interest.

A modern monograph on this topic is not known so far. There is only Lenschow's *Probing the Atmospheric Boundary Layer* from 1986. Books on general meteorological measurement methods like those by Brock and Richardson (2001), Strangeways (2003), or Emeis (2010) are designed to give a general overview of measurement techniques. The first two of these even devote only a few pages on ground-based remote sensing at all.

Therefore, up to now, a standard reference for ground-based remote sensing of the ABL is the review paper on "Surface-Based Remote Sensing of the Atmospheric Boundary Layer" by Wilczak et al. (1996) honouring the 25th anniversary of the scientific journal "Boundary-Layer Meteorology". A lot of progress has been made since the appearance of this review paper. Micro rain RADARs and the use of ceilometers for ABL profiling are completely new in ABL research and are not mentioned at all in Wilczak et al.'s paper. The same refers to modern small wind LIDARs for wind energy applications. ABL flux profile measurements, RASS techniques, and mixing-layer height determination are mentioned in Wilczak's review, but were at their infancy at that time. Considerable enhancements have been obtained since then.

This development of ground-based remote-sensing instrumentation in turn has revitalized the interest in boundary layer meteorology. Modern active and passive sounders allow for monitoring the whole depth of the boundary layer. The processing of this data partly needs advanced descriptions of the boundary layer, which could not be derived from pure surface layer probing. This above sketched development justifies a monograph on ground-based remote sensing of the atmospheric boundary layer.

Part of the material gathered in this book has been collected when preparing the author's lectures on modern meteorological measurement methods at the University of Cologne and at a summer school at Risø National Laboratory (part of DTU, Denmark) on "Remote sensing for Wind Energy". Some other parts are based on various peer-reviewed papers on ground-based remote sensing by the author, or those to which he contributed major parts. I am very grateful to Michael Hantel as member of the Editorial Board of this book series for bringing my attention to this platform for scientific communication. Following his suggestion, I read through a larger amount of recent scientific papers and tried to write an update on the present state of ground-based remote sensing of the ABL. I hope that practitioners and scientists will benefit from the collected material in this book. I thank the publisher for the interest in this important subject of current meteorological research and for the support and encouragement during the production of the book.

Garmisch-Partenkirchen, Germany Stefan Emeis

References

Brock FV, Richardson SJ (2001) Meteorological Measurement Systems. Oxford University Press, Oxford, 290 pp.

Emeis S (2010) Measurement Methods in Atmospheric Sciences – In situ and Remote. Borntraeger, Stuttgart, 257 pp.

Foken T (2008) Micrometeorology. Springer, New York

Kraus H (2008) Grundlagen der Grenzschicht-Meteorologie. Springer, New York

Lenschow D (1986) Probing the Atmospheric Boundary Layer. American Meteorological Society, Boston, MA, 269 pp.

Strangeways I (2003) Measuring the Natural Environment, second edn. Cambridge University Press, Cambridge, 534 pp.

Stull RB (1988) An Introduction to Boundary Layer Meteorology. Kluwer, Dordrecht

Wilczak JM, Gossard EE, Neff WD, Eberhard WL (1996) Ground-based remote sensing of the atmospheric boundary layer: 25 years of progress. Bound-Lay Meteor 78:321–349

Contents

List of Abbreviations

ABL	atmospheric boundary layer
ADMS urban	Atmospheric Dispersion Modelling System (Urban)
AERI	Atmospheric Emitted Radiance Interferometer
ARE	acoustic received echo (method)
ART	Algebraic Reconstruction Technique
BR	buoyancy ratio
CAPPI	constant altitude plane position indicator
CASES	Cooperative Atmosphere-Surface Exchange Study
CBL	convective boundary layer
CERC	Cambridge Environmental Research Consultants
CFL	constant flux layer
CNR	Consiglio Nazionale delle Ricerche
COST	European Cooperation in Science and Technology
CT	computed tomography
DBS	Doppler-beam-swinging
DGV	Doppler global velocimetry
DIAL	differential absorption LIDAR
DOAS	differential optical absorption spectroscopy
DTU	Danish Technical University
EARE	enhanced ARE
EISAR	Emission Infrared Spectrometer for Atmospheric Research
ESRL	Earth System Research Laboratory
FMCW	frequency-modulated continuous-wave
FORM	Foehn in the Rhine Valley during MAP
FTIR	Fourier transform infrared (absorption spectroscopy)
HATPRO	Humidity And Temperature PROfiler
HRDL	High-Resolution Doppler LIDAR
HSRL	high spectral resolution LIDAR
HWS	horizontal wind speed (method)
IBL	internal boundary layer
I-DOAS	imaging DOAS
IEC	International Electrotechnical Commission

IPSL	Institut Pierre Simon Laplace
IS	internal sublayer
ISAC	Institute of Atmospheric Sciences and Climate
ISARS	International Symposium on Acoustic Remote Sensing
ITU	International Telecommunications Union
LAS	large aperture scintillometer
LDA	laser Doppler anemometry
LDV	laser Doppler velocimetry
LED	Light-emitting diode
LIDAR	light detection and ranging
LLJ	low-level jet
LLLJP	Lamar Low-Level Jet Project
MAP	Mesoscale Alpine Programme
MAPS	multi-component air pollution software
MBL	Marine boundary layer
METEK	Meteorologische Messtechnik GmbH
MLH	Mixing-layer height
MO	Master oscillator
MOST	Monin–Obukhov similarity theory
MWS	micro-wave scintillometer
MWSR	microwave scanning radiometer
NOAA	National Oceanic and Atmospheric Administration
PBL	planetary boundary layer
PIV	particle imaging velocimetry
PO	Power oscillator
PPI	plane position indicator
RADAR	radio detection and ranging
RAMS	Regional Atmospheric Modeling System
RASS	radio-acoustic sounding system
RHI	range height indicator
RMSE	root mean square error
SAS	small aperture scintillometer
SBFM	Smooth Basis Function Minimization
SBL	stable boundary layer
SIGIS	scanning infrared gas imaging system
SIRT	Simultaneous Iterative Reconstruction Technique
SIRTA	Site Instrumental de Recherche par Télédétection Atmosphérique
SNODAR	Surface layer NOn-Doppler Acoustic Radar
SNR	Signal-to-noise ratio
SODAR	sound detection and ranging
SONAR	sound navigation and ranging
TDSI	time-dependent stochastic inversion
TIROS	television and infrared observation satellite
TUC	Temperature, hUmidity, and Cloud
UBL	urban boundary layer

UCL	urban canopy layer
UHF	ultra high frequency
URL	urban roughness layer
UV	ultraviolet
VAD	velocity azimuth display
VHF	very high frequency
VWV	vertical wind variance (method)
WAsP	Wind Atlas Analysis and Application Program
WPR	windprofiler

Chapter 1
Introduction

The atmospheric boundary layer (ABL) is that layer of the atmosphere which has direct influence on all living species on Earth. On the other hand, it is the layer which links the Earth's atmosphere to the remaining compartments of the system Earth. Therefore, a detailed knowledge on all features of this layer is mandatory for a physical understanding of the energy and mass fluxes and cycles in the system Earth and for the assessment of the living conditions in the biosphere. These energy and mass cycles in the system Earth are often addressed as biogeochemical cycles today. Thus, ABL research not only forms part of atmospheric sciences and meteorology but also has become an essential part of Earth system research and Global Change research.

The overall depth of the ABL can vary considerable from a few tens of metres to several kilometres. So, in any case, surface observations and measurements are only able to detect the state of the surface layer which forms the lowest part of the ABL. The larger part of the ABL cannot be probed directly without bringing in situ instruments into this part of the layer by mounting them on masts, balloons, or aircraft. The only other choice is indirect measurements by remote-sensing methods, either from the ground or from satellites. But it is only ground-based remote sensing that offers sufficient vertical resolution and temporal coverage for detailed boundary layer studies. Therefore, remote probing of the atmospheric boundary layer by ground-based remote sensing is the subject of this book.

1.1 The Beginnings of Boundary Layer Research

Specialized research on the atmospheric boundary layer started with surface and tower observations of vertical wind profiles and the transfer of laboratory experience on wall flows from fluid dynamics to meteorology in the first half of the twentieth century.

An unprecedented opportunity for probing the ABL away from the very surface became available about a hundred years ago, when wireless broadcasting emerged as a new medium in the first two decades of the twentieth century. The necessary infrastructure required the erection of large towers to disseminate the radio signals. Scientists rapidly detected these towers as a good platform to perform wind profile

S. Emeis, *Surface-Based Remote Sensing of the Atmospheric Boundary Layer*, Atmospheric and Oceanographic Sciences Library 40, DOI 10.1007/978-90-481-9340-0_1,

measurements (Hellmann 1915, Peppler 1921) and mounted cup anemometers in several heights at these towers. The new database made scientists think about the shape of the vertical wind profile. Hellmann and Peppler empirically derived from their measurements that the wind speed is proportional to the fourth root of the height above ground near the surface and to fifth root in higher levels between 16 and 124 m. Apart from this, both researchers noted that the diurnal variation of the wind speed differs between lower heights (having a daytime maximum in layers lower than about 50 m) and upper heights (exhibiting a nocturnal maximum).

Seminal theoretical papers, which originated shortly afterwards, were those of Prandtl (1925, 1932) where he introduced the mixing length concept for the description of boundary layer turbulence and for the first time formulated the logarithmic law for the vertical wind profile near the surface. Wind observations to heights of about 500 m were made at about the same time by Mildner (1932) in 1931, near Leipzig. He observed the ascents of 28 pilot balloons with two theodolites. His measurements became later famous as the "Leipzig wind profile" due to the evaluations of Lettau (1950) who re-examined this data set, and due to the formulation of the maximum mixing length in the ABL by Blackadar (1962), which is again based on Lettau's re-examination.

Similarly important were the papers of Obukhov (1946), Monin and Obukhov (1954), and Kazanski and Monin (1960), who introduced a similarity theory which is now known as the Monin Obukhov similarity theory (MOST). MOST describes the vertical profiles of wind and turbulent fluxes in the surface layer of the atmosphere, which is often called Prandtl layer. The validity of the MOST was investigated and tested in several field experiments in the 1960s and 1970s. The most famous experiments were the Wangara experiment in July and August 1967 over flat terrain mainly covered with sparse grass in Australia (Clarke et al. 1971) and the Kansas experiment in a wheat-farming country in summer 1968 in the United States (Izumi 1971). The evaluations of wind and flux data from mast measurements during these two experiments (16 m with the Wangara experiment and 32 m with the Kansas experiment) laid the foundations for the Prandtl-layer profile laws described in Chapter 2. Today, these profiles are commonly addressed as Dyer–Businger profiles (Businger et al. 1971, Dyer 1974).

The part of boundary layer meteorology that concentrates on flux measurements near the surface has become a special subject within boundary layer meteorology called micrometeorology (Foken 2008). Micrometeorology will be no subject in this book, because this book focuses on the overall vertical structure of the boundary layer. Therefore, we will address measurement techniques in the remaining sections of this introduction, which are suited to investigate and monitor the whole depth of the boundary layer.

1.2 Short History of Upper-Air Measurements

Until the beginning of the renaissance, people were just able to sense what was in their direct neighbourhood at the Earth surface. The only possibility to learn something about higher layers of the atmosphere would have been to climb up in the mountains, making observations on the slopes and at mountain summits. But, obviously, nobody seemed to be interested in such endeavours before the middle of the seventeenth century. The situation changed with the invention of the barometer by Evangelista Torricelli (1608–1647). Torricelli recognized that the column of mercury in his barometer was supported by the pressure of the atmosphere acting on the surface of the mercury in the dish, and in a letter to Michelangelo Ricci in 1644 he made the dramatic statement that "We live submerged at the bottom of an ocean of the element air, that by unquestioned experiments is known to have weight" (West 2005). If one leaves the bottom of this ocean, the weight of the remaining column above the observer should decrease. In 1648, Blaise Pascal (1623–1662) persuaded his brother-in-law, Florin Périer (1605–1672), to ascend the Puy de Dôme in France and to carry a barometer with him. He detected what was expected: that air pressure decreased with height; and from then on pressure readings were used to determine altitude. Fostered by new ideas coming up in the era of enlightenment, mountains became an object of interest for natural researchers in the following 30–40 years. For example, from 1774 on, Horace-Bénédict de Saussure (1740–1799) made temperature, humidity, and radiation measurements in the Alps. In 1781, the first mountain observatory on the Hohenpeissenberg, Germany, started its observations within the network of the Societas Meteorologica Palatina, which are continued without interruptions until today.

The next remarkable attempt to explore the atmosphere away from the surface was letting soar kites for scientific purposes (Fig. 1.1 shows an example from the beginning of the twentieth century). In 1749, the Scottish scientists Alexander Wilson (1714–1786) and Thomas Melville attached thermometers to kites and measured and compared temperatures at different altitudes above ground. Three years later, Benjamin Franklin (1706–1790) performed his famous kite ascents in order to investigate the electrical field in the air before the onset of thunderstorms.

Then, from 1783 on, climbing up mountains was no longer enough. Human beings began to travel up into the air with balloons, leaving the solid ground behind. Scientists immediately discovered this new opportunity for making meteorological

Fig. 1.1 Kite for atmospheric sounding from the beginning of the twentieth century at the Richard-Aßmann Observatory of the German Weather Service in Lindenberg (east of Berlin), Germany

measurements. The French professor of physics Jacques Charles (1746–1823) was the first. He used a hydrogen-filled balloon to reach a height of 2770 m on December 1, 1783. During this flight, he made temperature and pressure observations. Such balloon flights were resumed more systematically at the end of the nineteenth century. In 1894, Artur Berson (1859–1942) reached, with the help of inhaled oxygen, an altitude of 9155 m with a balloon. While the ascents at the end of the eighteenth century met a general curiosity, Berson's flights were driven by a real scientific interest in the vertical structure of the atmosphere.

After the inauguration of the meteorological observatory at Lindenberg, east of Berlin, Germany, in the year 1905, kite measurements were made there on a routine basis (Fig. 1.1). A special winch house, which was turnable in azimuthal direction, was erected for performing the kite ascents (Fig. 1.2). Until 1919, the world record for kite ascents was pushed up to 9750 m. About 10 years later, German and Russian meteorologists developed the radiosonde, a small box with basic meteorological instrumentation (temperature, relative humidity, pressure) that was carried by a balloon up into the stratosphere. Altitudes of 30 km were reached. Radiosondes

Fig. 1.2 Azimuthally turnable winch house at the Richard-Aßmann Observatory of the German Weather Service in Lindenberg (east of Berlin), Germany, built at the beginning of the twentieth century

Fig. 1.3 Balloon carrying an ozone sonde just after launching from the Hohenpeissenberg observatory of the German Weather Service

are still in use today and serve as a backbone of the worldwide meteorological upper-air network. Today, special radiosondes for the measurement of ozone concentration profiles (Fig. 1.3) are in use as well.

All these measurements mentioned above were directed to the exploration of the vertical structure of the atmosphere as a whole, without any special emphasis on the atmospheric boundary layer. This also happened – as mentioned earlier – because the recognition of the boundary layer as a separate scientific subject in meteorology did not start before the middle of the first half of the twentieth century.

1.3 The Beginning of Ground-Based Remote Sensing

Radiosondes are not suited to deliver continuous information in time, and they have a limited vertical resolution. Meteorological masts could serve for this purpose only for the first few hundreds of metres (e.g. the boundary meteorology mast at Obninsk southwest of Moscow and the mast at the Boulder Atmospheric Observatory in Colorado have a height of 300 m), which is by far not sufficient to monitor the full depth of the boundary layer. The only observation technique that can fill this gap is remote sensing.

Ground-based remote sensing started with the construction of the first RADAR devices. Originally developed for military purposes, first meteorological observations of moving rain bands were made in 1946 by R. Wexler and D.M. Swingle (Wexler and Swingle 1946). In the same year, the first acoustic remote-sensing devices, SODARs, were constructed by G.W. Gilman, H.B. Coxhead, and F.H. Willis (Gilman et al. 1946) and employed for the analysis of the vertical structure of the boundary layer. Only the backscatter intensity was recorded from the first RADAR and SODAR measurements (see, e.g. McAllister et al. (1969) for the description of the advantages of facsimile plots from acoustic sounding for the analysis of the vertical structure of the lower atmosphere). Computing resources for the analysis of the Doppler shift in order to detect wind speeds were missing at that time. The first Doppler RADAR was operated in 1957 (Brantley and Barczys 1957). In acoustic sounding, Doppler analysis to derive wind speeds was used from 1964 onwards (Kelton and Bricout 1964, Mahoney et al. 1973).

The next push came from the emerging laser techniques. Thus, in 1963, the first LIDARs (initially called optical RADAR) were used for meteorological observations (Fiocco and Smullin 1963). By 1970, scientists and engineers had conceived many LIDAR techniques and were building systems to demonstrate and apply this technology. LIDAR ceilometry (cloud-base height measurement) and observations of vertical aerosol structure, including inference of the inversion height, had both been well demonstrated. LIDAR observations of smokestack plume rise and of transport and dispersion of a cloud of aircraft-sprayed insecticide had also been performed. A differential absorption of light (DIAL) measurement of the vertical profile of water vapour was reported as early as 1964 (for a first review on DIAL, see Schotland 1974). Wind measurements achieved by tracking puffs of aerosol particles had been demonstrated, and Doppler measurements of wind motions were just beginning. However, practical applications of LIDAR were hampered by several difficulties, including fickle lasers, inadequate data systems, and eye safety restrictions. Experimenters were also encountering a number of problems, such as inadequate laser frequency control, interferences from other atmospheric constituents, and gaps in the theory of optical interaction with the atmosphere, all of which required solution before certain atmospheric parameters could be retrieved accurately and dependably (Wilczak et al. 1996).

The development of remote-sensing techniques was fostered by the appearance of the first satellites in space. Already 3 years after the first Russian satellite "Sputnik", the first meteorological satellite TIROS was launched in 1960. Spaceborne remote sensing, which employs optical or electromagnetic waves, has a limited vertical resolution, which is not sufficient for detailed boundary layer studies.

The investigation and monitoring of the boundary layer, therefore, had to rely on ground-based remote-sensing techniques. SODAR and RADAR are able to detect wind profiles and precipitation; LIDAR instruments basically give backscatter profiles. But the near-surface experiments (Wangara and Kansas, see above) had shown that temperature and humidity profiles are important as well. This finally led to the development of radio-acoustic sounding systems (RASS) and Raman-LIDAR systems. RASS determines vertical profiles of the speed of sound and from this the vertical temperature profile. The first attempts with such an instrument were made in 1972 (Marshall et al. 1972). First humidity profiling was made in 1970 with a Raman-LIDAR (Cooney 1970).

A first review of ground-based remote-sensing techniques for probing the clear atmosphere has been given by Derr and Little (1970); the further development of the sensor technology between 1970 and 1995 with special attention of the atmospheric boundary layer is described in Wilczak et al. (1996). The present state of ground-based remote-sensing instrumentation is summarized in Chapter 3.

1.4 The Scope of This Book

This book will concentrate on ground-based remote sensing of the atmospheric boundary layer, which has become a major tool for experimental meteorology in the past decades and which has seen considerable development since the last review by Wilczak et al. (1996). In situ and satellite observation techniques will only be addressed if they form an integral part of the presented ground-based remote-sensing technique.

Chapter 2 will gather some basic information on the vertical structure of atmospheric boundary layer over different surface types. Both major parts of the ABL, the surface or Prandtl layer and the Ekman layer, are considered. The most important analytical profile laws for wind and turbulence are given.

Chapter 3 introduces into the presently available instrumentation for ground-based remote sensing along horizontal and vertical paths in the ABL. The presented sounding techniques comprise acoustic, optical, and electromagnetic techniques.

Chapter 4 presents examples of knowledge on the vertical structure of different types of boundary layers, which has been obtained by ground-based remote sensing in recent years. This comprises mixing layer and boundary layer height as well as vertical wind, turbulence, temperature, humidity, and aerosol profiles. The analysis of these profiles does give information not only on the vertical structure of the boundary layer but also on secondary flow and circulation features occurring within the boundary layer.

The development and use of ground-based remote-sensing techniques still sees considerable progress. Therefore, a short outlook on possible future developments of ground-based remote sensing of the boundary layer will be given in Chapter 5.

Two appendices give a short overview of the series of ISARS conferences, which are devoted to boundary layer remote sensing and a glossary of special terms frequently used in remote sensing.

References

Blackadar AK (1962) The vertical distribution of wind and turbulent exchange in a neutral atmosphere. J Geophys Res 67:3095–3102
Brantley JQ, Barczys DA (1957) Some weather observations with a continuous-wave Doppler radar. Proceedings of the 6th Weather Radar Conference, American Meteorological Society, Cambridge, MA, pp 297–306
Businger JA, Wyngaard JC, Izumi Y, Bradley EF (1971) Flux profile relationships in the atmospheric surface layer. J Atmos Sci 28:181–189
Clarke RH, Dyer AJ, Brook RR, Reid DG, Troup AJ (1971) The Wangara experiment: boundary-layer data. Tech Paper No 19, Div Meteorol Phys CSIRO Australia, 21 pp.
Cooney, J (1970) Remote measurements of atmospheric water-vapour profiles using the Raman component of laser backscatter. J Appl Meteorol 9:182–184
Derr VE, Little CG (1970) A comparison of remote sensing of the clear atmosphere by optical, radio, and acoustic radar techniques. Appl Opt 9:1976–1992
Dyer AJ (1974) A review of flux-profile relations. Bound-Lay Meteorol 7:363–372
Fiocco G, Smullin LD (1963) Detection of scattering layers in the upper atmosphere /60-140 km/ by Optical RADAR. Nature 199:1275–1276
Foken T (2008) Micrometeorology. Springer, New York
Gilman GW, Coxhead HB, Willis FH (1946) Reflection of sound signals in the troposphere. J Acoust Soc Amer 18:274–283
Hellmann G (1915) Über die Bewegung der Luft in den unterste Schichten der Atmosphäre. Meteorol Z 34:273–285
Izumi Y (1971) Kansas 1968 Field Program Data Report. Air Force Cambridge Res Lab, AFCRL-72-0041 Environ Res Papers No 379, 79 pp.
Kazanski AB, Monin AS (1960) On turbulent conditions above the atmospheric surface layer. Izvestia Akad Nauk SSSR, Geophys Series No 1, 165–168
Kelton G, Bricout P (1964) Wind velocity measurements using sonic techniques. Bull Amer Meteorol Soc 45:571–580
Lettau, H (1950) A re-examination of the 'Leipzig Wind Profile' considering some relations between wind and turbulence in the frictional layer. Tellus 2:125–129
Mahoney AR, McAllister LG, Pollard JR (1973) The remote sensing wind velocity in the lower troposphere using an acoustic sounder. Bound-Lay Meteorol 4:155–167
Marshall JM, Peterson AM, Barnes AA (1972) Combined radar-acoustic sounding system. Appl Opt 11:108–112
McAllister LG, Pollard JR, Mahoney AR, Shaw PJR (1969) Acoustic sounding – A new approach to the study of atmospheric structure. Proceedings of the IEEE 57:579–587
Mildner P (1932) Über Reibung in einer speziellen Luftmasse. Beitr Phys fr Atmos 19:151–158
Monin AS, Obukhov AM (1954) Basic laws of turbulent mixing in the ground layer of the atmosphere. Trudy Geofiz Inst Akad Nauk SSSR No 24, 163–187
Obukhov AM (1946) Turbulence in an atmosphere with inhomogeneous temperature. Tr Inst Teor Geofiz Akad Nauk SSSR 1:95–115
Peppler, A (1921) Windmessungen auf dem Eilveser Funkenturm. Beitr Phys fr Atmos 9:114–129
Prandtl, L (1925) Bericht über Untersuchungen zur ausgebildeten Turbulenz. Zeits angew Math Mech 5:136–139
Prandtl, L (1932) Meteorologische Anwendungen der Stroemungslehre. Beitr Phys fr Atmos 19:188–202
Schotland RM (1974) Errors in the lidar measurement of atmospheric gases by differential absorption. J Appl Meteor 13:71–77
West JB (2005) Robert Boyle's landmark book of 1660 with the first experiments on rarified air. J Appl Physiol 98:31–39
Wexler R, Swingle DM (1946) Radar storm detection. Bull Amer Meteor Soc 28:159–167
Wilczak JM, Gossard EE, Neff WD, Eberhard WL (1996) Ground-based remote sensing of the atmospheric boundary layer: 25 years of progress. Bound-Lay Meteorol 78:321–349

Chapter 2
Analytical Description and Vertical Structure of the Atmospheric Boundary Layer

2.1 Introduction

The atmospheric boundary layer (ABL) is the lower part of the troposphere, and by this the lowest layer of the atmosphere as a whole. In contrast to the free atmosphere above, the presence of the underlying Earth's surface has a measurable influence on the ABL. It is the only part of the atmosphere where frictional forces play an important role, and where the temperature can exhibit a diurnal variation. The ABL acts as a kind of broker that communicates the transport of energy, momentum, and other substances between the Earth surface and the free atmosphere, i.e. the ABL is dominated by vertical fluxes of these variables. These fluxes have their largest amounts directly at the surface and vanish at the top of the ABL.

This chapter does not offer a complete overview on the ABL, but rather concentrates on the vertical structure of the ABL. More general descriptions of the ABL can be found in Stull (1988), Arya (1995), Garratt (1992), and other books. Because the ABL enwraps the whole Earth, it is often also called planetary boundary layer (PBL).

The Earth's surface is a place where turbulence is generated; it is usually a sink for atmospheric momentum, and either a source or a sink for heat, moisture, and other atmospheric compounds such as aerosols and trace gases. Therefore, we find in the ABL less momentum but more turbulence, and different heat, moisture, aerosol, and pollutant concentrations than in the free atmosphere above. The detection of the vertical profiles of the just mentioned atmospheric variables can thus help to identify the vertical structure and extend of the ABL. This detection of these profiles and their interpretation is the main topic of this book.

Three principal types of the ABL can be distinguished: (i) if heat input from below dominates, we find a convective boundary layer (CBL); (ii) if the atmosphere is cooled from below, we find a stable boundary layer (SBL); and (iii) if the heat flux at the lower surface is vanishing and dynamical shear forces are dominating, we find a neutral or dynamical boundary layer. We will start with the principal description of the vertical structure of these three ABL types in the subsection on flat terrain below.

The vertical structure of these three ABL types additionally depends to a large extent on the type and texture of the underlying surface. Its shape, roughness,

S. Emeis, *Surface-Based Remote Sensing of the Atmospheric Boundary Layer*, Atmospheric and Oceanographic Sciences Library 40, DOI 10.1007/978-90-481-9340-0_2,

albedo, moisture content, heat emissivity, and heat capacity determine the momentum and energy exchange between the surface and the atmosphere. The vertical extend of the ABL types is mainly determined by the generation of turbulent kinetic energy at and the input of heat from the lower surface. The then following subsections will present some of the most important formings of the ABL types with respect to the surface characteristics like the urban boundary layer (UBL) or the marine boundary layer (MBL). In pure specification, these formings will only appear if the flow is in equilibrium with the underlying surface. Each time when the horizontal atmospheric flow crosses a boundary from one surface type or subtype to the next, a new internal boundary layer forms, which will eventually – if no further change in surface conditions takes place – reach a new equilibrium.

2.2 ABL Over Flat Terrain

The simplest structure of the ABL is found over flat, horizontally homogeneous terrain with uniform soil type and land use and a uniform distribution of roughness elements. Its vertical stratification in roughness sublayer, constant-flux sublayer (Prandtl layer), and Ekman layer is depicted in Fig. 2.1. This flat-terrain ABL is mainly coined by the diurnal variation of the energy balance of the Earth's surface. During daytime, when the sun is heating the ground, a CBL is growing due to the input of heat from below, which generates thermal convection. During night-time, when the ground cools due to the emission of long-wave radiation, a new nocturnal SBL forms near the ground (see Fig. 2.2). If clouds, wind, and precipitation override

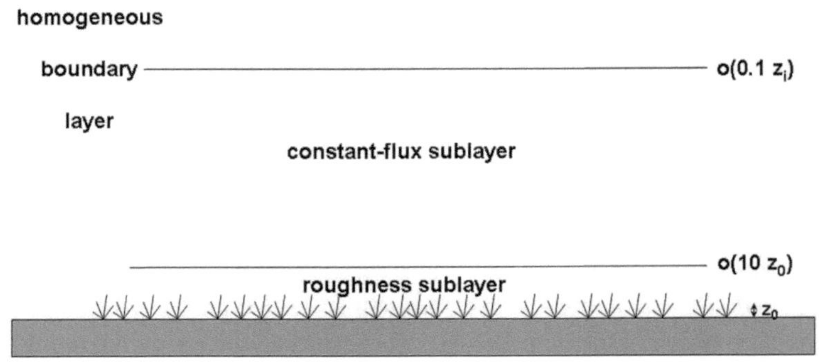

Fig. 2.1 Schematic diagram of the atmospheric boundary layer over flat homogeneous terrain. Heights are given in terms of the roughness length, z_0, and the boundary layer height, z_i

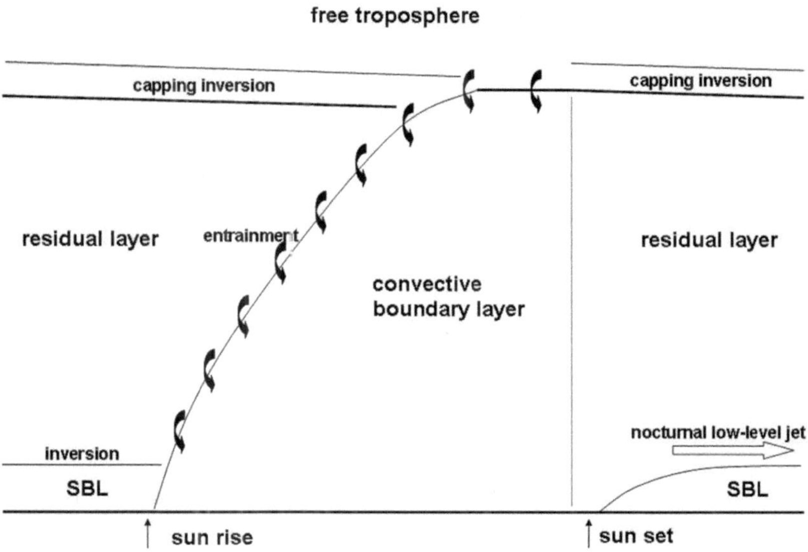

Fig. 2.2 Schematic diagram of the diurnal variation of the atmospheric boundary layer from midnight to midnight (adapted from Stull 1988). SBL designates the nocturnal stable boundary layer

the influence of short-wave and long-wave radiation, the ABL is even simpler and a neutral boundary layer with nearly no diurnal variation forms. Its depth is then mainly determined by the magnitude of the wind shear within it and by the advection of warmer or colder air masses aloft with a prescribed thermal stratification.

2.2.1 Neutral (Dynamical) Boundary Layer

The neutral ABL is usually found with high wind speeds and in weather situations with clouds which prohibit heating by incoming solar radiation or cooling by outgoing long-wave radiation. Apart from a viscous or laminar sublayer directly above the surface that is only a few millimetres deep, we have two main compartments of this ABL type: (i) the Prandtl layer or constant-flux layer and (ii) the Ekman layer. The equilibrium conditions in the neutral boundary layer are described by a balance between a driving large-scale pressure gradient force, the Coriolis force due to the rotation of the Earth, and the dissipating frictional force. The horizontal equations of motions for this equilibrium read:

$$\frac{1}{\rho}\frac{\partial p}{\partial x} + fv + \frac{\partial}{\partial z}K_{\mathrm{M}}\frac{\partial u}{\partial z} = f(v - v_{\mathrm{g}}) + \frac{\partial}{\partial z}K_{\mathrm{M}}\frac{\partial u}{\partial z} = 0, \quad (2.1)$$

$$\frac{1}{\rho}\frac{\partial p}{\partial y} - fu + \frac{\partial}{\partial z}K_{\mathrm{M}}\frac{\partial v}{\partial z} = -f(u - u_{\mathrm{g}}) + \frac{\partial}{\partial z}K_{\mathrm{M}}\frac{\partial v}{\partial z} = 0, \quad (2.2)$$

where p is the atmospheric pressure, ρ is the atmospheric density, f is the Coriolis parameter, u is the wind component in x-direction, v is the wind component in y-direction, z is the vertical coordinate, and K_M is the coefficient for vertical turbulent momentum exchange. The subscript "g" denotes the respective components of the geostrophic wind, theoretically blowing in a perfect balance between pressure gradient force and Coriolis force. The introduction of the geostrophic wind eliminates the pressure gradient from the equations of motions (2.1) and (2.2). $K_M \left(\partial u/\partial z\right)$ is a parameterization of the vertical turbulent momentum flux.

2.2.1.1 Prandtl Layer

The Prandtl layer or constant-flux layer is defined as that layer where the turbulent vertical fluxes of momentum, heat, and moisture deviate less than 10% from their surface values, and where the influence of the Coriolis force is negligible. Usually, this layer covers only 10% of the whole ABL depth. Although this definition seems to be paradox because the turbulent vertical fluxes have their largest vertical gradients just at the surface, the concept of the constant-flux layer has proven to be a powerful tool to describe the properties of this layer. Stipulating a vertically constant momentum flux and assuming that the mean flow is in x-direction simplifies the equations of motion (2.1) and (2.2) further to

$$K_M \frac{\partial u}{\partial z} = \text{const} = u_*^2, \tag{2.3}$$

where u_* is the friction velocity. The friction velocity is proportional to the geostrophic wind speed and thus represents the large-scale pressure gradient force. Dynamical considerations, which suggest formulating the exchange coefficient as being proportional to the mixing length $l = \kappa z$, which in turn is proportional to the distance to the ground, and the friction velocity ($K_M = \kappa u_* z$), lead to the following equation for the vertical wind gradient in the Prandtl layer (with the van Kármán constant $\kappa = 0.4$):

$$\frac{\partial u}{\partial z} = \frac{u_*}{l} = \frac{u_*}{\kappa z}. \tag{2.4}$$

Integration of eq. (2.4) from a lower height z_0 where the wind speed vanishes to a height z within the Prandtl layer yields the well-known logarithmic wind profile with the roughness length z_0:

$$u(z) = \frac{u_*}{\kappa} \ln \frac{z}{z_0}. \tag{2.5}$$

Sometimes, instead of eq. (2.5), an empirical power law is used to describe the vertical wind profile:

$$u(z) = u(z_r) \left(\frac{z}{z_r}\right)^a, \tag{2.5a}$$

where z_r is a reference height and a is the power law exponent (sometimes called "Hellmann exponent"). a depends on surface roughness and the thermal stability of the Prandtl layer. (A comparison of eqs. (2.5) and (2.5a) is discussed in Emeis (2005).) The wind speed increases with height without a turning of the wind direction. A scale analysis gives for the height of the Prandtl layer z_p (Kraus 2008):

$$z_p \approx 0.01\frac{u_*}{f} \approx 0.00064\frac{|v_g|}{f}. \tag{2.6}$$

Putting in numbers ($u_* = 0.5$ m/s, $f = 0.0001\ 1/s, v_g = 8$ m/s) gives a typical height of the Prandtl layer of 50 m.

In a well-mixed Prandtl layer, the temperature T is decreasing with height according to the adiabatic lapse rate, g/c_p (g is gravity acceleration, c_p is specific heat at constant pressure). The potential temperature, $\Theta = T(p_0/p)^{R/}c_p$ is constant with height (R is the gas constant for dry air).

The standard deviations of the three velocity components (u denotes the streamwise component, v crosswise, w vertical) are independent of height and scale with the friction velocity (Stull 1988, Arya 1995):

$$\frac{\sigma_u}{u_*} \approx 2.5; \quad \frac{\sigma_v}{u_*} \approx 1.9; \quad \frac{\sigma_w}{u_*} \approx 1.3. \tag{2.7}$$

The turbulence intensity in the streamwise direction decreases with height due to the increase of the mean wind speed with height. By inserting the left-hand relation from eq. (2.7) into eq. (2.5), we get (Wieringa 1973)

$$\frac{\sigma_u}{u(z)} = \frac{1}{\ln(z/z_0)}. \tag{2.8}$$

2.2.1.2 Ekman Layer

The Ekman layer covers the major part of the ABL above the Prandtl layer. If the simplifying assumption is made that the height-dependent growth of the exchange coefficient K_M stops at the top of the Prandtl layer and that it is vertically constant within the Ekman layer, we can solve eqs. (2.1) and (2.2) analytically for the wind profile in the Ekman layer:

$$u(z) = u_g(1 - \exp(-\gamma z)\cos\gamma z), \tag{2.9}$$

$$v(z) = u_g \exp(-\gamma z)\sin\gamma z, \tag{2.10}$$

where $\gamma = \sqrt{f/(2K_M)}$. Equation (2.9) describes the well-known "Ekman spiral." The height of the Ekman layer is estimated by

$$z_E = \frac{\pi}{\gamma} \tag{2.11}$$

2.2.1.3 Unified Description of the Wind Profile for the Boundary Layer

For many purposes, a unified description of the wind profile for the lower part of the ABL is desirable, which is valid beyond the surface layer. Due to the assumption of the constant exchange coefficient K_M, the relations (2.9) and (2.10) cannot be extended from the Ekman layer down into the Prandtl layer. Likewise, due to the assumption of a mixing length, which grows linearly with height, the relation (2.5) cannot be extended into the Ekman layer. Therefore, two approaches have been tested to overcome this problem. The first idea is to fit the Prandtl and Ekman profiles in such a way together that there is a smooth transition in terms of wind speed and wind shear between both regimes. The second idea is to modify the mixing length in order to extrapolate the Prandtl layer wind profile into higher layers.

Etling (2002) has proposed the first idea by presenting a wind profile description with a linearly increasing K_M below the Prandtl layer height, z_p, and a constant K_M above this height:

$$u(z) = \begin{cases} u_*/\kappa \ln(z/z_0) & \text{for } z < z_p \\ u_g(-\sin\alpha_0 + \cos\alpha_0) & \text{for } z = z_p \\ \begin{aligned} &u_g[1 - 2\sqrt{2}\exp(-\gamma(z - z_p)) \\ &\sin\alpha_0 \cos(\gamma(z - z_p) + \pi/4 - \alpha_0) \\ &+2\exp(-2\gamma(z - z_p))\sin^2\alpha_0]^{1/2} \end{aligned} & \text{for } z > z_p \end{cases} \tag{2.12}$$

Equation (2.12) depends on five parameters: the surface roughness z_0, the geostrophic wind speed u_g, the height of the Prandtl layer z_p, the friction velocity u_*, and the angle between the surface wind and the geostrophic wind α_0. The two variables z_0 and u_g are external parameters, the other three of them are internal parameters of the boundary layer. If a fixed value is chosen for z_p, then two further equations are needed to determine u_* and α_0. Equation (2.12) describes a smooth transition of wind speed from the Prandtl layer to the Ekman layer.

Deviating from Etling (2002), these equations should be generated from the more realistic physical requirement that both the wind speed as well as the wind shear are continuous in the height $z = z_p$ (Emeis et al. 2007). Equating the first two equations of the wind profile, equation (2.12) for $z = z_p$ gives

$$u_* = \frac{\kappa u_g(-\sin\alpha_0 + \cos\alpha_0)}{\ln(z_p/z_0)}, \tag{2.13}$$

and from equating the respective shear equations in the same height $z = z_p$, we get a second equation for u_*

$$u_* = 2\left|u_g\right|\gamma\kappa z_p \sin\alpha_0. \tag{2.14}$$

Equating now the right-hand sides of eqs. (2.13) and (2.14) yields the desired relation for α_0

$$\alpha_0 = arctg \frac{1}{1 + 2\gamma z_p \ln(z_p/z_0)}. \tag{2.15}$$

Equation (2.15) still depends on the friction velocity u_* via the definition of γ:

$$\gamma = \sqrt{\frac{f}{2\kappa u_* z_p}}. \tag{2.16}$$

Thus, u_* has to be determined iteratively starting with a first guess for u_* in eq. (2.16), subsequently computing α_0 from eq. (2.15), and then re-computing u_* from eqs. (2.13) or (2.14). Inversely, the system of eqs. (2.13) to (2.16) can be used to determine z_p if u_* is known. Application of eq. (2.12) needs the knowledge of three internal parameters.

The second idea, to modify the dependence of the mixing length with height, has been proposed by Gryning et al. (2007). They reformulated the mixing length l ($= \kappa z = L_L$ in the Prandtl layer) in order to limit its growth with height and thus to extend the validity of eq. (2.5) to above the surface layer. They choose

$$\frac{1}{l} = \frac{1}{L_L} + \frac{1}{L_M} + \frac{1}{L_U}. \tag{2.17}$$

A modified mixing length is formed in eq. (2.17) by introducing a length scale for the middle part of the boundary layer, $L_M = u_*/f(-2 \ \ln(u_*/fz_0)) + 55)^{-1}$ and a length scale for the upper part, $L_U = (z_i - z)$. This results in the following wind profile alternative to eq. (2.5) or eq. (2.12):

$$u(z) = \frac{u_*}{\kappa}\left(\ln\frac{z}{z_0} + \frac{z}{L_M} - \frac{z}{z_i}\frac{z}{2L_M}\right). \tag{2.18}$$

Peña et al. (2009) suggest a similar approach starting from Blackadar's (1962) principal approach for the mixing length, l

$$l = \frac{\kappa z}{1 + \left(\frac{\kappa z}{\eta}\right)^d}, \tag{2.19}$$

which can be rewritten as

$$\frac{1}{l} = \frac{1}{\kappa z} + \frac{(\kappa z)^{d-1}}{\eta^d}, \tag{2.20}$$

and incorporated this approach into the logarithmic profile law giving

$$u(z) = \frac{u_*}{\kappa}\left(\ln\frac{z}{z_0} + \frac{1}{d}\left(\frac{\kappa z}{\eta}\right)^d - \frac{1}{1+d}\frac{z}{z_i}\left(\frac{\kappa z}{\eta}\right)^d - \frac{z}{z_i}\right). \tag{2.21}$$

For neutral stability and $d = 1$, Peña et al. (2009) find $\eta = 39$ m; for $d = 1.25$ they give $\eta = 37$ m. The only necessary parameter in eq. (2.21) from above the surface layer is the height of the boundary layer, z_i.

2.2.1.4 Internal Boundary Layers

The boundary layer flow structure tends to be in equilibrium with the surface properties underneath, which govern the vertical turbulent momentum, heat, and moisture fluxes. When the flow transits from one surface type to another with different surface properties, the flow structure has to adapt to the new surface characteristics. This leads to the formation of an internal boundary layer (IBL, internal because it is a process taking place within an existing boundary layer) that grows with the distance from the transition line (Fig. 2.3).

An IBL with a changed dynamical structure can develop when the flow enters an area with a different roughness (e.g. from pasture to forests or from agricultural areas to urban areas). An IBL with a modified thermal structure can come into existence when the flow enters an area with a different surface temperature (e.g. from land to sea or from water to ice). Often dynamical and thermal changes occur simultaneously. Vertical profiles of wind, temperature, and moisture can change at the IBL top.

The IBL top has to be distinguished from inversions and sloping frontal surfaces at which likewise changes in the vertical profiles of wind, temperature, and moisture can happen. Inversions are usually horizontal and caused either by adiabatic sinking

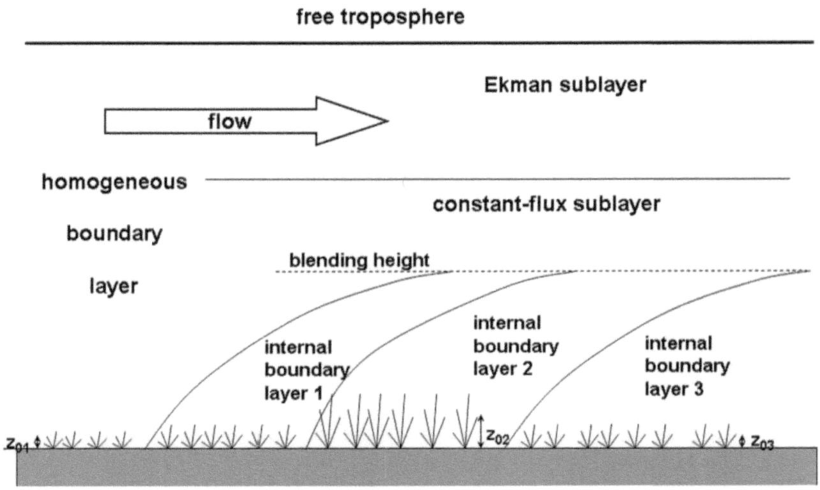

Fig. 2.3 Formation of internal boundary layers due to changes in surface roughness length

motions from above or by radiative cooling from below. Frontal surfaces are slanted like IBL tops, but they are moving with the synoptic pressure systems and are not linked to changes in surface properties.

If several subsequent changes in the surface properties occur in streamwise direction multiple IBL can form. They all grow with distance from the respective boundary in surface type. In some larger distance to the initiating change in surface properties, the single IBL looses its identity and multiple IBLs can no longer be distinguished from each other. This height is also called blending height.

2.2.2 Convective Boundary Layer

The thermal stratification of the ABL is rarely found to be really neutral. In most cases, there is a non-vanishing virtual potential heat flux $\overline{\Theta'_v w'}$ at the ground (the virtual potential temperature, Θ_v includes the modifying influence of the atmospheric humidity on the static stability of the air), which leads to a thermal stratification of the ABL. From this surface heat flux and the friction velocity u_*, a length scale, L_*, the Obukhov length, can be formed,

$$L_* = \frac{\Theta}{\kappa g} \frac{u_*^3}{\overline{\Theta'_v w'}}. \tag{2.22}$$

The heat flux is counted positive if it is directed from the atmosphere towards the ground (cooling the atmosphere) and negative if it is towards the atmosphere (heating the atmosphere). Thus an unstable ABL is characterized by a negative Obukhov length. The virtual potential heat flux $\overline{\Theta'_v w'}$ can be separated into a heat flux and a humidity flux,

$$\overline{\Theta'_v w'} = \overline{\Theta' w'} + 0.61\Theta \overline{q' w'}. \tag{2.23}$$

The ratio of the turbulent heat flux and the humidity flux is called Bowen ratio,

$$B = \frac{c_p \overline{\Theta' w'}}{L_v \overline{q' w'}}, \tag{2.24}$$

where q is specific humidity. The buoyancy exerted by the vertical heat and humidity gradients is given by $\frac{g}{\Theta}\overline{\Theta' w'} + 0.61\, g\overline{q' w'}$. The ratio of these two contributing terms is called the buoyancy ratio, BR, which is inversely proportional to the Bowen ratio:

$$\mathrm{BR} = \frac{0.61\Theta \overline{q' w'}}{\overline{\Theta' w'}} = \frac{0.61 c_p \Theta}{L_v} \frac{1}{B}. \tag{2.25}$$

This unstable type of the ABL is usually found during daytime over surfaces heated by insolation and over waters that are warmer than the air above. The degree of instability is described by the non-dimensional ratio of the height z over the

Obukhov length L_*. For small negative values of z/L_*, the vertical profiles in the CBL can be described by introducing a correction function $\Psi_m(z/L_*)$ (Paulson 1970, Högström 1988),

$$\Psi_m = 2\ln\left(\frac{1+x}{2}\right) + \ln\left(\frac{1+x^2}{2}\right) - 2arctg(x) + \frac{\pi}{2}, \tag{2.26}$$

where $x = (1 - b\ \ z/L_*)^{1/4}$ and b=16. With this correction function, the wind profile (2.12) becomes

$$u(z) = \begin{cases} u_*/\kappa\ (\ln(z/z_0) - \Psi_m(z/L_*)) \quad \text{for } z < z_p \\ u_g(-\sin\alpha_0 + \cos\alpha_0) \quad \text{for } z = z_p \\ u_g[1 - 2\sqrt{2}\exp(-\gamma(z - z_p)) \\ \sin\alpha_0 \cos(\gamma(z - z_p) + \pi/4 - \alpha_0) \quad \text{for } z > z_p \\ +2\exp(-2\gamma(z - z_p))\sin^2\alpha_0]^{1/2} \end{cases} \tag{2.27}$$

In the unstable case, eqs. (2.13) to (2.15) take the following forms:

$$\sigma_{u,v,w} = 0.6 w_*, \tag{2.28}$$

$$u_* = \frac{2\left|u_g\right| \gamma\kappa z_p \sin\alpha_0}{\phi(z_p/L_*)}, \tag{2.29}$$

$$\alpha_0 = arctg\frac{1}{1 + \frac{2\gamma z_p}{\phi(z_p/L_*)}\left(\ln(z_p/z_0) - \Psi_m(z_p/L_*)\right)}, \tag{2.30}$$

where ϕ is the differential form of the correction function Ψ for thermal stratification:

$$\varphi(z/L_*) = (1 + b\ z/L_*)^{-1/4}\,. \tag{2.31}$$

u_* and α must be determined in the same iterative procedure as described after eq. (2.16). γ still has the form given in eq. (2.16), b is set to 16 following Högström (1988).

The alternative approaches by Gryning et al. (2007) and Peña et al. (2009) yield the following wind profiles, which could be used in place of eq. (2.27),

$$u(z) = \frac{u_*}{\kappa}\left(\ln\frac{z}{z_0} + T(\frac{z}{L_*}) + \frac{z}{L_M} - \frac{z}{z_i}\frac{z}{2L_M}\right), \tag{2.32}$$

$$u(z) = \frac{u_*}{\kappa}\left(\ln\frac{z}{z_0} - \Psi_m + \frac{1}{d}\left(\frac{\kappa z}{\eta}\right)^d - \frac{1}{1+d}\frac{z}{z_i}\left(\frac{\kappa z}{\eta}\right)^d - \frac{z}{z_i}\right), \tag{2.33}$$

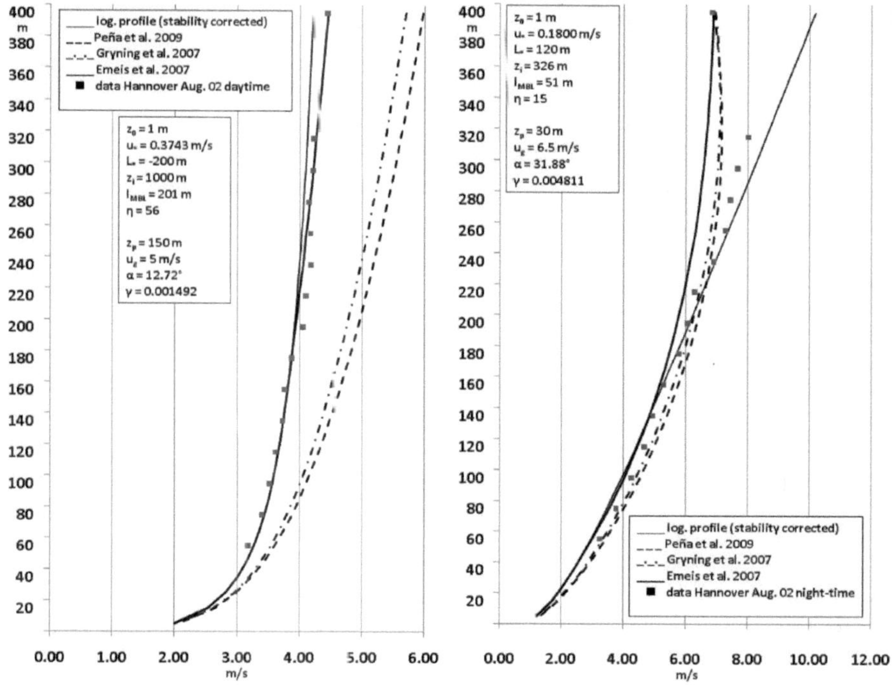

Fig. 2.4 Comparison of different wind profile descriptions with measured sodar data for daytime unstable conditions (*left*) and night-time stable stratification (*right*). The *thin full lines* give the logarithmic profile with stability correction (upper eq. in (2.27)), the *bold full line* the profile from the lower eq. in (2.27). The *dashed profile* is computed from eq. (2.33) and the *dash-dotted profile* from eq. (2.32)

with a complex stability correction function $T(z/L_*)$ and Ψ_m from eq. (2.26). For details the reader is referred to Gryning et al. (2007) and Peña et al. (2009). A comparison of wind profiles from eqs. (2.27), (2.32), and (2.33) is shown in Fig. 2.4.

Temperature is decreasing with height at least by the adiabatic lapse rate, in a shallow super-adiabatic layer near the surface even stronger. The standard deviations of the wind components in the Prandtl layer (below z_p) now depend on the ABL height z_i or the height z (Panofsky et al. 1977, Arya 1995):

$$\frac{\sigma_{u,v}}{u_*} = \left(12 - 0.5\frac{z_i}{L_*}\right)^{1/3}, \tag{2.34}$$

$$\frac{\sigma_w}{u_*} = 1.3\left(1 - 3\frac{z}{L_*}\right)^{1/3}. \tag{2.35}$$

For the Ekman layer above z_p, Arya (1995) gives

$$\sigma_{u,v,w} = 0.6 w_*, \tag{2.36}$$

with the convective velocity scale,

$$w_* = \frac{g z_i}{\Theta} \overline{w'\Theta'}. \tag{2.37}$$

This means that the standard deviation of the vertical velocity component increases with height in the unstable Prandtl layer and then stay constant above it.

2.2.3 Stable Boundary Layer

This type of ABL, which is characterized by an downward (positive) surface heat flux ($L_* > 0$) and a stable thermal stratification of the air, is usually found at night-time, over waters that are colder than the air above, and over ice and snow-covered surfaces. For positive values of z/L_*, the correction function for eq. (2.26) reads (Businger et al. 1971, Dyer 1974, Högström 1988, Lange and Focken 2006)

$$\Psi_m(z/L_*) = \begin{cases} -az/L_* & \text{for } 0 < z/L_* \le 0.5 \\ Az/L_* + B(z/L_* - C/D) \\ \exp(-Dz/L_*) + BC/D & \text{for } 0.5 \le z/L_* \le 7 \end{cases}, \tag{2.38}$$

where $a = 5, A = 1, B = 2/3, C = 5$, and $D = 0.35$. Then the wind profile $u(z)$, u_*, and α can be computed again from eqs. (2.27) to (2.30), with

$$\varphi(z/L_*) = 1 + a\, z/L_*. \tag{2.39}$$

The temperature in the stable boundary layer vertically decreases less than the adiabatic lapse rate; the potential temperature increases with height. The standard deviations of the wind components are usually assumed to be constant with height in the same way as described by eq. (2.7) for the neutral ABL (Arya 1995).

Under suitable weather conditions (anticyclonic, sufficient large-scale pressure gradient, clear skies), a low-level jet can form at the top of a nocturnal SBL. When the atmosphere stabilizes rapidly after sunset, the vertical turbulent momentum flux decreases rapidly and the wind-braking effect of the rough surface is no longer communicated to the air at the top of the nocturnal SBL. This rapid change in the equilibrium of forces leads to an inertial oscillation. Prominent features of this oscillation are a speed-up of the wind speed in the first half of the night and a turning wind direction throughout the night. The phenomenon ends with the restoration of the vertical turbulent momentum flux after sunrise. Following Jacobi and Roth (1995), this inertial oscillation of the wind components is described by:

$$\frac{\partial u}{\partial t} = -f(v_g - v), \tag{2.40}$$

$$\frac{\partial v}{\partial t} = f(u_g - u). \tag{2.41}$$

The solution of this system of equations is given by

$$u - u_g = D_v \sin(ft) + D_u \cos(ft), \tag{2.42}$$

$$v - v_g = D_v \cos(ft) - D_u \sin(ft). \tag{2.43}$$

The differences $(u_g - u)$ and $(v_g - v)$ are called ageostrophic wind components, and D_u and D_v denote the ageostrophic components at the onset of the inertial oscillation.

2.3 Urban Boundary Layers

Urban agglomerations have recently received special interest in atmospheric boundary layer studies. Nowadays, more than half of mankind is living in cities and the number of megacities with more than 10 million inhabitants is steadily growing. Cities are large pollution sources and because the temperature is already higher than in their surroundings, they are especially prone to the effects of a warming climate. All these aspects have fostered studies on the structure of the UBL. UBL meteorology has become a special subject in boundary layer meteorology. One aspect of UBL studies is the analysis of wind profiles and thermally driven secondary circulations over cities (urban heat islands). See for an overview of urban meteorology and of urban heat islands, e.g. Kanda (2007) and Hidalgo et al. (2008). The urban heat island brings about a secondary circulation with winds towards the urban centre near the ground, uprising motion over the urban centre and compensating outflow towards the surrounding rural areas aloft (Shreffler 1978, 1979).

Urban surfaces are characterized by large roughness elements, wide-spread sealed areas, reduced moisture availability at the surface, and increased possibilities for heat storage. This leads to higher turbulence intensities in the UBL and to stronger sensible heat fluxes from the urban surface into the UBL. Both facts induce a greater depth of the boundary layer (see the urban dome in Fig. 2.5). Additionally, towns are often isolated islands featuring these special surface properties surrounded by rural terrain so that the flow above them is not in equilibrium with the urban surface. During daytime the reduced moisture availability leads to smaller latent and thus larger sensible heat fluxes at the urban surface compared to rural surfaces. The reduced radiative cooling of the urban surface or even still persisting upward heat fluxes (Velasco et al. 2007) at night prevents the formation of a stable nocturnal boundary layer. Both the increased sensible heat flux during the day and the reduced cooling during the night cause higher temperatures in the UBL compared to the surrounding rural boundary layer. This effect is known as the urban heat island (Atkinson 2003, Chow and Roth 2006). The urban heat island is enhanced by the human energy production (Crutzen 2004, Kanda 2007), which with 20–70 Wm^{-2} can be 5–10% of the energy input by solar irradiation.

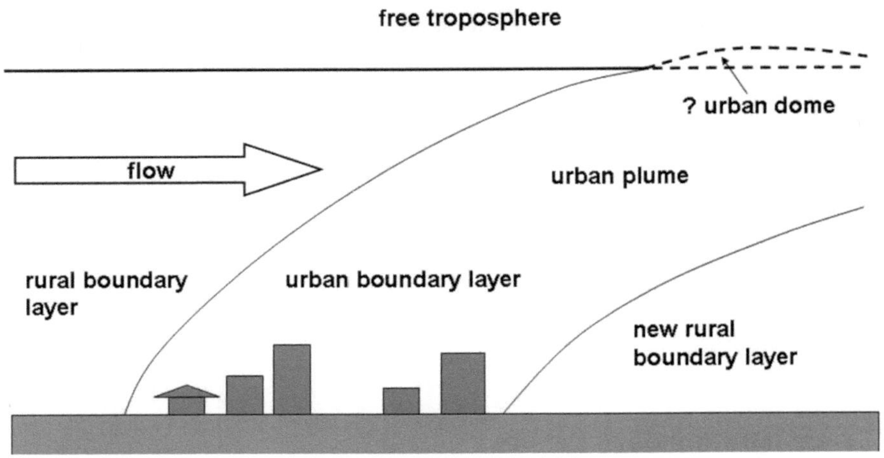

Fig. 2.5 Urban plume downwind of a larger city. This is a special case of an internal boundary layer (cf. Fig. 2.3)

In a horizontal flow, the presence of the city is a change in surface properties. Following Section 2.2.1.4, this leads to the formation of internal boundary layers (Fig. 2.5). The internal layer formed by the properties of the urban surface is often called urban plume.

Following Plate (1995), Roth (2000), and Piringer et al. (2007), the urban boundary layer (UBL) is usually divided into four layers (Fig. 2.6). The lowest one

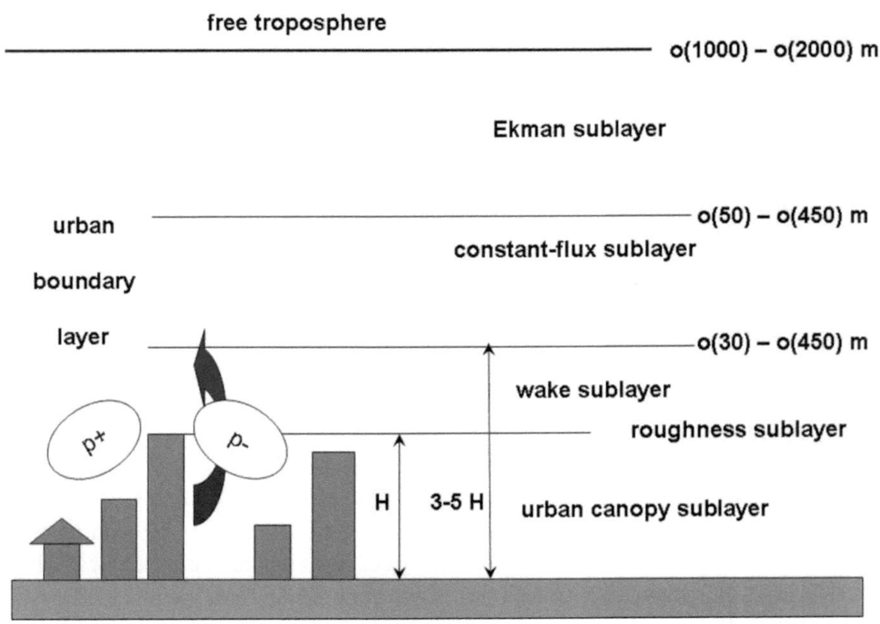

Fig. 2.6 Vertical layering in an urban boundary layer. H gives the average building height, $p+$ and $p-$ designate atmospheric pressure disturbances upstream and downstream of single buildings

is the urban canopy layer (UCL), which reaches up to the mean top height of the buildings. The next layer is the wake layer in which the influence of single buildings on the flow is still notable. This wake layer usually extends to about two to five times the average building height. These two layers are often jointly addressed as the urban roughness sublayer (URL, Rotach 1999). Strong vertical exchange by forced vertical motions can occur in this layer. Above the urban roughness layer is the constant flux layer (CFL) or inertial sublayer (IS), over homogeneous terrain usually addressed as surface layer or Prandtl layer. In the uppermost part of the boundary layer above the CFL, the wind direction turns into the direction of the geostrophic wind (often called Ekman layer). If a convectively driven boundary layer (CBL) is present, no distinction is made between the CFL or Prandtl-layer and the Ekman layer, but they are jointly addressed as mixing layer. Good overviews of the special features of the UBL can, for example, be found in Roth (2000), Arnfield (2003), and Grimmond (2006).

Wind and turbulence within the UBL determine the horizontal and vertical dispersion and transport of air pollutants in towns and is thus important for the health of the citizens. Numerous field experiments (for an overview see, e.g. Grimmond (2006)), numerical studies (see, e.g. Batchvarova and Gryning (2006)), and several wind tunnel studies (Counihan 1973, Farell and Iyengar 1999, Schatzmann and Leitl 2002) therefore have been conducted to investigate the structure of the UBL. Besides a better understanding of mixing and transport processes within the UBL, a realistic representation of the flow field within street canyons and above the buildings is essential for the application of dispersion models to urban areas (e.g. model simulations for London with ADMS Urban (CERC 2001)).

2.4 Forest Boundary Layers

Forest-covered surfaces are a special form of vegetated surfaces. The special features of the forest boundary layer decisively depend on the density of the trees. If trees grow very close together, their crowns form a rough surface, which has much in common with a rough grass land (Raupach 1979) as depicted in Fig. 2.1. But if the trees grow sparser, then the surface has many similarities with an urban surface and must be described as shown in Fig. 2.7. In contrast to the urban canopy layer (Fig. 2.6), the forest canopy layer must be subdivided in two layers: the stem layer and the crown layer. In the stem sublayer the horizontal wind speed may be higher than in the denser crown sublayer. The main difference between a densely vegetated forest (Fig. 2.1) and a sparsely vegetated forest (Fig. 2.7) is that larger air parcels can enter (sweeps) and leave (ejections) the forest canopy sublayer. This permeability of the forest canopy sublayer leads to an anomaly featuring higher turbulence intensities in the wake sublayer than expected from the mean vertical wind gradient in this layer. Therefore, the usual flux-gradient relationships are not valid in the whole roughness sublayer (Högström et al. 1989). This anomalous wake layer may extend to about three to five tree heights.

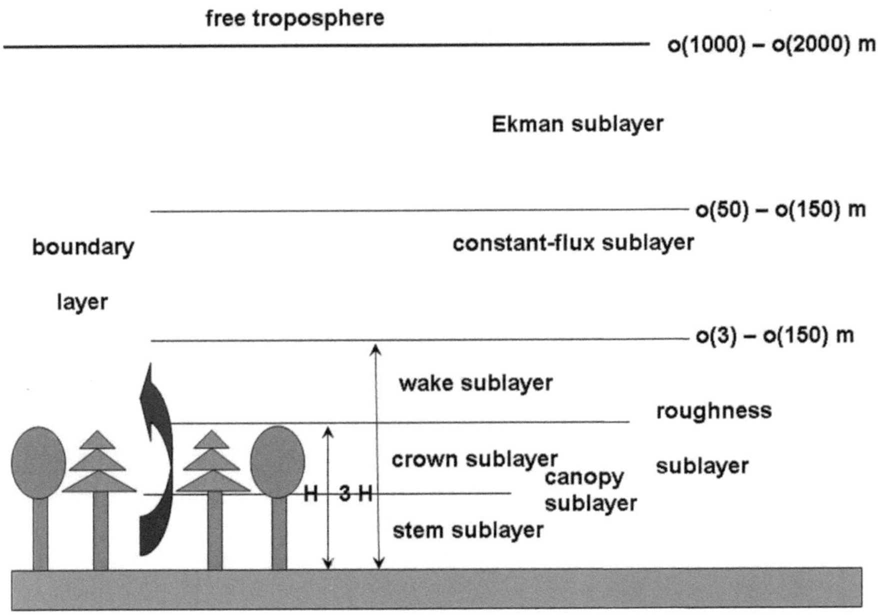

Fig. 2.7 Vertical layering in a forest boundary layer. *H* gives the average tree height

An overview of the turbulence in the plant canopy sublayer and the air layer just above it is given by Finnigan (2000). A more general overview is compiled in Finnigan (2007).

2.5 Marine Boundary Layers

In contrast to all other boundary layer types, the marine boundary layer exhibits special features due to the physical properties of the sea surface such as waves, the large heat capacity of the water, and the large humidity transfer from the water to the lower atmosphere.

The flexibility of the wavy water surface allows for a two-way interaction between the air and the water. The wind can modify the sea surface by the formation of waves, and existing waves in turn can modify the air flow above them. This leads to a complicated interaction between the wind fields in the marine boundary layer and the wave fields on the ocean surface. The basic cycle of this interaction can be described as follows. When the wind speed increases ahead of an approaching atmospheric depression, waves (young waves in the sense of eq. (2.44) below) are generated by frictional forces (tangential stress) and pressure forces (form drag) (Banner and Peirson 1998) and grow as long as the downward momentum flux u_*^2 in the air is larger than the potential energy gh of the existing waves with wave height h. The build-up of the waves is the more effective the longer the area is within which the wind can act on the sea surface (the so-called fetch), and it takes some time to reach an equilibrium state between the wind field and the wave field. Therefore, the maximum wave height is reached usually several hours after the peak wind speed (Emeis and Türk 2009). When the wind wanes, the existing waves have a large inertia and continue to propagate. Thus, on the rear side of a decaying depression, the wave speed can easily be larger than the wind speed. Now the waves (old waves in the sense of eq. (2.44)) can drive the air at the lower boundary of the marine boundary layer, and the sea surface is no longer a sink but a source for atmospheric momentum.

The large thermal inertia of the water leads to cycles and time scales in the temperature difference between the surface and the air that are not known from any other terrestrial surfaces. The temperature of the ocean water is dominated by an annual cycle while a diurnal cycle is near non-existing. This cycle is delayed one to two months to the annual cycle of the air temperature. Therefore, we usually find water warmer than the air and an unstably stratified marine boundary layer in late summer, autumn, and early winter, while colder waters than the air and a stable stratification of the marine boundary layer prevail in late winter, spring, and early summer. This mean temperature difference between ocean and atmosphere is modified by cold-air and warm-air advections coupled to moving atmospheric pressure systems (depressions and anticyclones). If the atmospheric pressure systems do not move too fast, warm-air advection or cold-air advection over a given part of the ocean surface can persist for a day or two. This can be considerably longer than the 6–12 hours for which a given thermal stratification can prevail over land.

The large humidity source of the water frequently leads to upward turbulent humidity fluxes into the usually unsaturated air above. Due to this, a very low Bowen ratio (eq. 2.24) of about 0.1 (Sempreviva and Gryning 1996) is observed in the marine boundary layer. This leads to a buoyancy ratio (see eq. (2.25)) of about 0.4, i.e. 30% of the buoyancy of air parcels in the marine boundary layer is produced by humidity fluxes. Edson et al. (2004) even say that the humidity flux component provides more than half of the total buoyancy flux. Thus, the marine surface layer is often slightly unstable although the potential temperature slightly increases with height.

While the thermal conditions of the marine boundary layer can be described with the same parameters that suffice for a description of onshore boundary layers, the dynamical interaction between the wave and the wind field at least requires the introduction of two additional parameters: the wave height, h and the wave age, ξ. Additionally the water depth, d, has some influence. Waves become higher but slower when they come to shallower waters near the coast because they are modified from deep-water waves to shallow-water waves. The wave age is the ratio between the wave speed c_{ph} of the dominant waves and either the wind speed or the friction velocity, u_*. Usually, the wave age is given by (Oost et al. 2000, 2002)

$$\xi = \frac{c_{ph}}{u_*}. \tag{2.44}$$

Employing definition (2.44), $\xi < 28$ denotes young (growing) waves and $\xi > 28$ old waves (swells).

The roughness of the sea surface, z_0, varies with the sea state. For a wave-free surface (low wind speeds, very small fetch), the sea surface is aerodynamically smooth (Banner and Peirson 1998) and follows the law for smooth wall turbulent boundary layers (where ν is the viscosity of the air),

$$z_0 = \frac{1}{11}\frac{\nu}{u_*}. \tag{2.45}$$

For a wavy surface, the roughness length is proportional to the wave height h. Therefore, in an equilibrium state, where the frictional force of the wind on the waves and the potential energy of the waves balance each other, we find a constant ratio

$$\frac{gz_0}{u_*^2} = A, \tag{2.46}$$

with the Charnock coefficient (or Charnock number) A to be determined from experimental data. Charnock himself gave approximately 0.015 (Charnock 1955). This is valid for a wave age (eq. 2.44) between 10 and 25 for wind speeds around 25–30 m/s. For higher winds speeds and growing seas the Charnock number increases with the wave age, for smaller wind speeds it decreases with wave age (Hara et al. 2004). For older waves ($\xi > 35$), Jones and Toba (2001) give a constant Charnock number of 0.008. Foreman and Emeis (2010) suggest 0.018 from a reanalysis of a large number of literature studies.

In the vertical the marine boundary layer can be subdivided into three layers: the wave sublayer within which the pressure disturbance by the single wave (which is responsible for the wave form drag) is observable, the constant-flux sublayer, and the Ekman sublayer (Fig. 2.8). Over a smooth wave-free sea surface, the wave

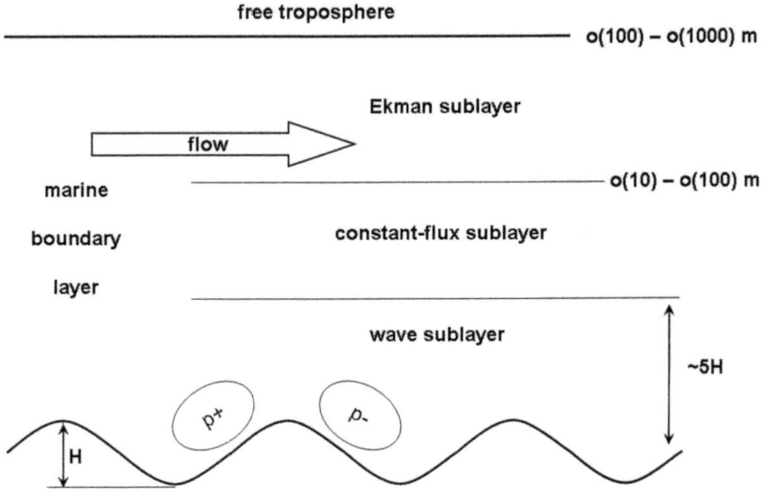

Fig. 2.8 Vertical layering in a marine boundary layer. H gives the wave amplitude, $p+$ and $p-$ designate atmospheric pressure disturbances upstream and downstream of single waves

sublayer is missing and is replaced by a viscous sublayer. The wave sublayer is approximately five wave heights deep and the constant-flux sublayer depth depends on the surface roughness, too.

At the top of unstably stratified marine boundary layers rolls and cellular convection patterns can develop. Since the detection and monitoring of these cloud features is beyond the present abilities of ground-based remote sensing, they are not addressed further here.

Overviews of air–sea interaction, the drag forces exerted on the marine boundary layer, and detailed discussions of momentum transfer between the sea surface and the air above may be found in Donelan (1990), Banner and Peirson (1998), and Foreman and Emeis (2010). Sullivan and McWilliams (2010) discuss the coupling processes between surface gravity waves and adjacent winds and currents in the turbulent boundary layers of the atmosphere and the ocean.

2.5.1 Land–Sea Wind System

Due to the different thermal inertia of land and sea surfaces, secondary circulation systems – land–sea wind systems – can form at the shores of oceans and larger lakes, which modify the ABL structure. Under clear-sky conditions and low to moderate winds, land surfaces become cooler than the adjacent water surface due to long-wave emittance at night and they become warmer than the water surface due to the absorption of short-wave irradiance during daytime. As a consequence, rising motion occurs over the warmer and sinking motion over the cooler surfaces. A flow from the cool surface towards the warm surface develops near the surface and a

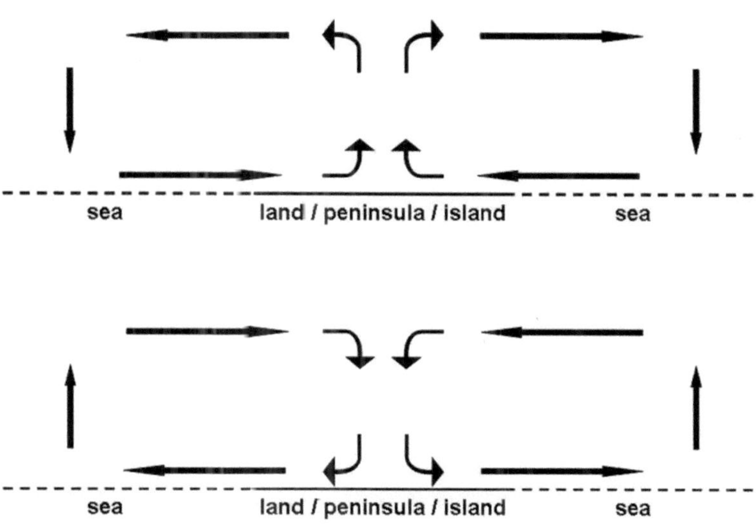

Fig. 2.9 Schematic presentation of a height cross-section of a land–sea breeze system during daytime (*top*) and night-time (*below*)

return flow emerges in the opposite direction aloft in order to keep the mass balance. This leads to the well-known sea breeze during daytime and in the evening and the land breeze at night and in the early morning (Fig. 2.9). This common feature of a land breeze is the reason why in former times sailing ships left harbours in the early morning and tended to return in the afternoon. This thermally driven secondary motion system can be several hundreds of metres deep. The sea breeze front propagates inland during the day and is – if enough moisture is available in the air – often marked by a chain of cumulus clouds.

2.6 Mountainous Terrain

The presence of hills and mountains lead to much larger horizontal inhomogeneities in the ABL than all that has been presented so far in this chapter. Larger mountains can even have a larger vertical extension than the depth of the ABL (Fig. 2.10). Thus, the applicability of eqs (2.1) to (2.43) can only be expected in small limited parts of the mountain boundary layer over smaller hills or in wide valleys with flat floor. The differences to the boundary layers introduced above come due to both mechanical and thermal forcing. While the mechanical forcing such as channelling of flows in valleys and large-scale blocking by mountain chains are quite obvious, the thermal forcing is more difficult to understand. The thermal forcing is a mixture of the presence of elevated heating (or cooling at night) surfaces and the reduced

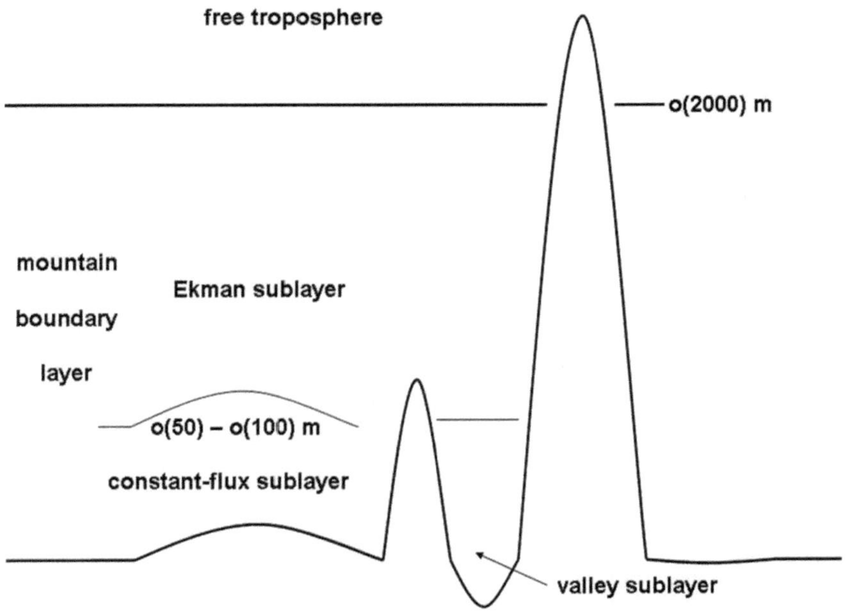

Fig. 2.10 Some features of the atmospheric boundary layer in mountainous terrain

ratio of the effected air volume to the thermally active surface area in mountainous terrain. As this book concentrates on boundary layer aspects, gravity wave and Foehn generation in thermally stably stratified flows over mountains will not be addressed here. The reader is rather referred to overview papers on these large-scale effects of mountain ranges, e.g. the classical one by Smith (1978).

Mountains lead to three types of thermally driven secondary circulation systems, which modify the vertical structure of the mountainous ABL: slope winds, mountain and valley winds, and – like land–sea wind systems – a diurnally changing system of winds between mountain ranges and the surrounding plains (see Fig. 2.11, which is an extension of the classical sketch from Defant (1949) that depicted only the first two of these three secondary circulation systems). These three phenomena occur on three different spatial scales although they all three have the same temporal scale of one day. Slope winds (thin arrows in Fig. 2.11) develop on a slope spatial scale of a few metres up to about 1 km. Mountain and valley winds (*full arrows* in Fig. 2.11) emerge on a spatial scale of a few hundred metres up to a few hundred kilometres in long valleys. Mountain-plain winds have the largest scale of a few tens of kilometres to more than one hundred kilometres (open arrows in Fig. 2.11).

Slope winds come into existence due to the heating by insolation or radiative cooling of a sloping surface in mountainous terrain. These winds emerge and disappear within minutes after the thermal forcing started or stopped. They form part of a secondary circulation in a valley cross section. Upslope winds during daytime may lead to compensating sinking motion over the centre of the valley (Vergeiner 1982). This is often the reason why clouds resolve over the valley centre, but form over hill crests. This sinking motion contributes to a stabilization of the thermal stratification in the valley atmosphere and can prolong the existence of temperature inversions in valleys. At night-time downslope winds develop.

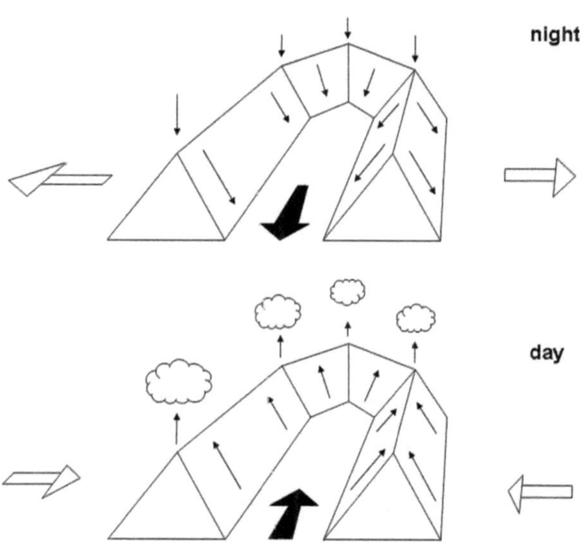

Fig. 2.11 Diurnal variation of local and regional wind systems in and around mountain ranges for night-time (*upper frame*) and daytime (*lower frame*). *Thin arrows*: upslope and downslope winds, *black arrows*: upvalley and downvalley winds, *white arrows*: regional winds between the mountain range and the surroundings plains (mountain pumping). Upslope winds during daytime support the formation of clouds over summits and crests

Mountain and valley winds take a few hours to form. They are a feature of the whole valley (Vergeiner and Dreiseitl 1987). Mountain winds (sometimes called down-valley winds, a better term would be out-valley winds because local slopes of the valley floor are not decisive (Heimann et al. 2007)) usually start 3–4 hours after sunset and valley winds (sometimes called up-valley winds or better in-valley winds) 3–4 hours after sun rise. Both winds require clear-sky conditions so that heating by incoming short-wave radiation and cooling by outgoing long-wave radiation can occur. The direction of the winds along a valley axis is dominated by the fact that heating and cooling of the valley air is more effective in the narrower upper parts of the valley than in the wider lower parts, because the ratio of air mass to thermally active surface is larger in the narrower upper parts of a valley (Steinacker 1984). This differential heating or cooling along the valley axis leads to a pressure gradient along the valley axis, which in turn drives the winds. Usually, the daytime in-valley winds are stronger and more turbulent than the nocturnal out-valley winds.

The large-scale wind system between a mountain range and the surrounding planes has the largest similarity with the land–sea wind system (Fig. 2.9). This wind system, which blows towards the mountains during daytime and away from the mountains at night-time, takes 4–6 hours to develop. It is observable even 100 km away from the foothills of a mountain range (Lugauer and Winkler 2005). This wind system comes into existence because at a given height above sea level, the air over the mountains is heated more than over the plane. The opposite occurs at night-time. This differential heating once again leads to a pressure difference at a given height, and this pressure difference in turn drives a compensating wind.

There must be a compensating wind system for the mountain and valley winds and for the mountain-plain winds as well. Because this compensating motion takes place over a larger area, it is too weak to be differentiated from the synoptic-scale motions. During daytime this compensating motion contributes to downward motions aloft over the surrounding plains of a mountain range that somewhat limits the vertical growth of clouds at the boundary layer top over these plains. For such a circulation system in Southern Germany, the term Alpine Pumping has been proposed (Lugauer and Winkler 2005).

References

Arnfield AJ (2003) Two decades of urban climate research: a review of turbulence, exchanges of energy and water, and the urban heat island. Int J Climatol 23:1–26

Arya SP (1995) Atmospheric boundary layer and its parameterization. In: Cermak JE et al. (eds) Wind Climate in Cities. Kluwer, Dordrecht, 41–66

Atkinson BW (2003) Numerical modelling of urban heat-island intensity. Bound-Lay Meteorol 109:285–310

Banner ML, Peirson WL (1998) Tangential stress beneath wind-driven air-water surfaces. J Fluid Mech 364:115–145

Batchvarova E, Gryning E-E (2006) Progress in urban dispersion studies. Theor Appl Climatol 84:57–67

Blackadar AK (1962) The vertical distribution of wind and turbulent exchange in a neutral atmosphere. J Geophys Res 67:3095–3102

Businger JA, Wyngaard JC, Izumi Y, Bradley EF (1971) Flux profile relationships in the atmospheric surface layer. J Atmos 28:181–189

CERC (2001) Cambridge Environmental Research Consultants, ADMS dispersion model, http://www.cerc.co.uk.

Charnock H (1955) Wind stress on a water surface. Quart J Roy Meteor Soc 81:639–640

Counihan J (1973) Simulation of an adiabatic urban boundary layer in a wind tunnel. Atmos Environ 7:673–689

Chow WTL, Roth M (2006) Temporal dynamics of the urban heat island of Singapore. Int J Climatol 26:2243–2260

Crutzen PJ (2004) New directions: The growing urban heat and pollution "island" effect – impact on chemistry and climate. Atmos Environ 38:3539–3540

Defant F (1949) Zur Theorie der Hangwinde, nebst Bemerkungen zur Theorie der Berg- und Talwinde. Arch Meteorol Geophys Bioklimatol A 1:421–450

Donelan M (1990) Air-sea interaction. In: LeMehaute B, Hanes DM (eds) The Sea. Wiley-Interscience, Hoboken, NJ, 239–292

Dyer AJ (1974) A review of flux-profile relations. Bound-Lay Meteorol 1:363–372

Edson JB, Zappa CJ, Ware JA McGillis WR, Hare JE (2004) Scalar flux profile relationships over the open ocean. J Geophys Res 109:C08S09. doi: 10.1029/2003JC001960

Emeis S (2005) How well does a power law fit to a diabatic boundary-layer wind profile? DEWI Mag 26:59–62

Emeis S, Türk M (2009) Wind-driven wave heights in the German Bight. Ocean Dyn 59:463–475

Emeis S, Baumann-Stanzer K, Piringer M, Kallistratova MA, Kouznetsov R, Yushkov V (2007) Wind and turbulence in the urban boundary layer – analysis from acoustic remote sensing data and fit to analytical relations. Meteorol Z 16:393–406

Etling D (2002) Theoretische meteorologie. Springer, Heidelberg etc., 354 pp.

Farell C, Iyengar AKS (1999) Experiments on the wind tunnel simulation of atmospheric boundary layers. J Wind Eng Indust Aerodyn 79:11–35

Finnigan J (2000) Turbulence in plant canopies. Ann Rev Fluid Mech 32:519–571

Finnigan JJ (2007) The turbulent wind in plant and forest canopies. In: Johnson EA, Miyanishi K (eds) Plant Disturbance Ecology. Academic Press, New york, 15–58

Foreman R, Emeis S (2010) Revisiting the definition of the drag coefficient in the marine atmospheric boundary layer. J Phys Oceanogr, in print, DOI 10.1175/2010JPO4420.1

Garratt JR (1992) The Atmospheric Boundary Layer. Cambridge University Press, Cambridge (UK)

Grimmond CSB (2006) Progress in measuring and observing the urban atmosphere. Theor Appl Climatol 84:3–22

Gryning S-E, Batchvarova E, Brümmer B, Jørgensen HE, Larsen S (2007) On the extension of the wind profile over homogeneous terrain beyond the surface layer. Bound-Lay Meteorol 124:251–268

Hara T, Belcher SE, Ginis I, Moon I-J (2004) Air-Sea Momentum Flux at High Winds. Meeting Amer Meteor Soc 2004, paper 2A.2, 2 pp (http://www.po.gso.uri.edu/Numerical/tropcyc/flux/amsmeet2004a.pdf)

Heimann D, De Franceschi M, Emeis S, Lercher P, Seibert P (eds) (2007) Air pollution, traffic noise and rekated health effects in the Alpine space – a guide for authorities and consulters. ALPNAP comprehensive report. Università degli Studi di Trento, Trento, 335 pp.

Hidalgo J, Masson V, Baklanov A, Pigeon G, Gimeno L (2008) Advances in urban climate modeling. Ann NY Acad Sci 1146:354–374

Högström U (1988) Non-dimensional wind and temperature profiles in the atmospheric surface layer: A re-evaluation. Bound-Lay Meteorol 42:55–78

Högström U, Bergström H, Smedman A-S, Halldin S, Lindroth A (1989) Turbulent exchange above a pine forest, I: Fluxes and gradients. Bound-Lay Meteorol 49:197–217

Jacobi C, Roth R (1995) Organisierte mesoskalige Störungen in der stabilen planetaren Grenzschicht. Meteorol Z NF 4:150–161

Jones ISF, Toba Y (2001) Wind Stress over the Ocean. Cambridge University Press, Cambridge (UK)

Kanda M (2007) Progress in urban meteorology: A review. J Meteor Soc Jap 85B:363–383

Kraus H (2008) Grundlagen der Grenzschicht-Meteorologie. Springer, New York

Lange M, Focken U (2006) Physical approach to short-term wind power prediction. Springer, Berlin, Heidelberg, etc., 208 pp.

Lugauer M, Winkler P (2005) Thermal circulation in South Bavaria – climatology and synoptic aspects. Meteorol Z 14:15–30

Oost WA, Jacobs CMJ, van Oort C (2000) Stability effects on heat and moisture fluxes at sea. Bound-Lay Meteorol 95:271–302

Oost WA, Komen GJ, Jacobs CMJ, van Oort C (2002) New evidence for a relation between wind stress and wave age from measurements during ASGAMAGE. Bound-Lay Meteorol 103:409–438

Panofsky HA, Tennekes H, Lenschow DH, Wyngaard JC (1977) The characteristics of turbulent velocity components in the surface layer under convective conditions. Bound-Lay Meteorol 11: 355–361

Paulson CA (1970) The mathematical representation of wind speed and temperature profiles in the unstable atmospheric surface layer. J Appl Meteorol 9:857–861

Peña A, Gryning S-E, Hasager CB (2009) Comparing mixing-length models of the diabatic wind profile over homogeneous terrain. Theor Appl Climatol 100:325–335

Piringer M, Joffre S, Baklanov A, Christen A, Deserti M, de Ridder K, Emeis S, Mestayer P, Tombrou M, Middleton D, Baumann-Stanzer K, Dandou A, Karppinen A, Burzynski J (2007) The surface energy balance and the mixing height in urban areas – activities and recommendations of COSTAction 715. Bound-Layer Meteor 124:3–24

Plate EJ (1995) Urban climates and urban climate modelling: An introduction. In: Cermak JE et al. (eds) Wind Climate in Cities. NATA ASI Series E277, Kluwer, Dordrecht, 23–39.

Raupach MR (1979) Anomalies in flux-gradient relationships over forest. Bound-Lay Meteorol 16:467–486

Rotach MW (1999) On the influence of the urban roughness sublayer on turbulence and dispersion. Atmos Environ 33:4001–4008

Roth M (2000) Review of atmospheric turbulence over cities. Quart J Roy Meteor Soc 126:941–990

Schatzmann M, Leitl B (2002) Validation and application of obstacle-resolving urban dispersion models. Atmos Environ 36:4811–4821

Sempreviva A, Gryning S-E (1996) Humidity fluctuations in the marine boundary layer measured at a coastal site with an infrared humidity sensor. Bound-Lay Meteorol 77:331–352

Shreffler JH (1978) Detection of centripetal heat – Island circulations from tower data in St. Louis. Bound-Lay Meteorol 15:229–242

Shreffler JH (1979) Heat island convergence in St. Louis during calm periods. J Appl Meteorol 18:1512–1520

Smith RB (1978) The influence of mountains on the atmosphere. In: Landsberg HE, Saltzman B (eds) Advances in Geophysics 21:87–230

Steinacker R (1984) Area-height distribution of a valley and its relation to the valley wind. Contr Atmos Phys 57:64–71

Stull RB (1988) An Introduction to Boundary Layer Meteorology. Kluwer, Dordrecht

Sullivan PP, McWilliams JC (2010) Dynamics of winds and currents coupled to surface waves. Ann Rev Fluid Mech 42:19–42. doi: 10.1146/annurev-fluid-121108-145541

Velasco E, Márquez C, Bueno E, Bernabé RM, Sánchez A, Fentanes O, Wöhrnschimmel H, Cárdenas B, Kamilla A, Wakamatsu S, Molina LT (2007) Vertical distribution of ozone and VOCs in the low boundary layer of Mexico City. Atmos Chem Phys Discuss 7:12751–12779

Vergeiner I (1982) An energetic theory of slope winds. Meteorol Atmos Phys 19:189–191

Vergeiner I, Dreiseitl E (1987) Valley winds and slope winds – observations and elementary thoughts. Meteorol Atmos Phys 36:264–286

Wieringa J (1973) Gust factors over open water and built-up country. Bound-Lay Meteorol 3:424–441

Chapter 3
Basic Principles of Surface-Based Remote Sensing

3.1 Introduction

In situ measurements at the ground or from masts, tethered balloons, radiosondes, or aircraft offer only limited access to the atmospheric boundary layer. They are restricted either to certain sites or to short time intervals or both. The wish to overcome these limitations has led to the development of remote-sensing techniques that allow the continuous probing of the whole depth of the atmospheric boundary layer with sufficient vertical resolution. Two main lines of development have been followed in the past 60 years for this purpose: active and passive ground-based remote sensing. Observations from satellites do not offer a sufficient vertical resolution for retrievals from the atmospheric boundary layer.

Most of the ground-based remote-sensing techniques presently used in boundary layer meteorology are active techniques, which are based on the controlled emission of some radiation and the analysis of a small backscattered fraction of this radiation. We will concentrate on these techniques here. A general overview of all meteorological measurement techniques can be found in Emeis (2010).

There are three relevant wavelength bands within which active remote sensing of the atmosphere based on backscatter is possible: centimetre to metre waves backscattered from refractive index gradients due to atmospheric turbulence, millimetre to centimetre waves backscattered from falling or soaring water droplets, and micrometer waves backscattered from aerosol particles and gas molecules (Fig. 3.1). In any case optimal backscatter is received if the scattering objects in the atmosphere have about half the size of the wavelength λ of the emitted optical, electromagnetic, or acoustic radiation (Bragg condition). The Bragg condition that is essential for an optimal atmospheric remote sensing has been described for the first time in 1913 by the British physicists William Henry Bragg (1862–1942) and William Lawrence Bragg (1890–1971), father and son, a few years after returning from a more than 20 year stay in Adelaide, Australia. They referred to the scattering of x-rays at crystals (Bragg and Bragg 1913). If the Bragg condition is met, backscatters from different centres of action have a maximum positive interference.

Further we have to differentiate between elastic (Mie and Rayleigh) and inelastic (Raman) scattering. Mie scattering occurs if the size of the scattering particles (molecules, aerosol particles, droplets, insects) has about the same wavelength as

S. Emeis, *Surface-Based Remote Sensing of the Atmospheric Boundary Layer*, Atmospheric and Oceanographic Sciences Library 40, DOI 10.1007/978-90-481-9340-0_3,

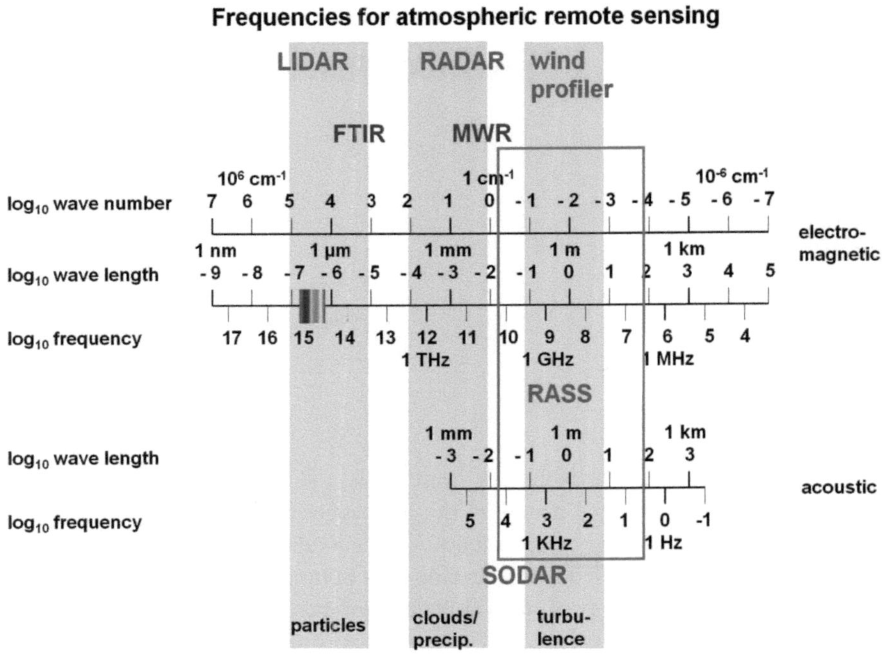

Fig. 3.1 Possible frequency ranges for remote sensing of the atmosphere. Selected active and passive remote sensing methods are marked (LIDAR, FTIR, etc.). Ceilometers operate in the same frequency range as LIDARs. The frame marks the range shown in more detail in Fig. 3.6

the radiation; Rayleigh scattering occurs if the size of the scattering object is much smaller than the wavelength of the radiation. Mie scattering has only small wavelength dependence, because the ratio between particle size and wavelength is close to the Bragg condition. Rayleigh scattering is proportional to λ^{-4}, i.e. the backscatter intensity increases drastically when coming closer to the Bragg condition. In the case of Raman scattering the energy of the incoming radiation is contributing temporarily to the rotation or vibration energy of a molecule and is then emitted again at a different frequency.

A second condition for a successful sounding of the atmosphere is that the atmosphere is not or only weakly absorbing or attenuating the used radiation in order to allow for a sufficient range of the remote-sensing technique. Wavelength bands within which the atmosphere is not absorbing are called atmospheric windows. One window is in the range of visible light, another in the infrared range at 10–15 μm. For radiation with shorter wavelengths than 0.3 μm (UV radiation), for some wavelengths between 0.9 and 10 μm, and for wavelengths greater than 15 μm, the atmosphere is more or less opaque. For large wavelength above about 1 mm (micro wave range) the atmosphere becomes transparent again.

The combination of relevant backscatter processes in the atmosphere and the wavelength bands within which the atmosphere is transparent has led to the development of the different active remote-sensing techniques used today (see Table 3.1

Table 3.1 Overview of the main ground-based boundary layer remote-sensing techniques

Name	Principle	Spatial resolution	Direction	Type
RADAR	Backscatter, electromagnetic pulses, fixed wavelength	Profiling	Scanning,slanted	Active, monostatic
SODAR	Backscatter, acoustic pulses, fixed wavelength	Profiling	Fixed, slanted, vertical	Active,usually monostatic
LIDAR	Backscatter, optical pulses, fixed wavelength(s)	Profiling	Scanning, fixed, horizontal, slanted, vertical	Active, monostatic
RASS	Backscatter, acoustic, electromagnetic, fixed wavelength	Profiling	Fixed, vertical	Active, monostatic
FTIR	Absorption, infrared, spectrum	Path-averaging	Fixed, horizontal, slanted	Active, bistatic or passive
	Emission, infrared, spectrum	Path-averaging	Fixed, horizontal, slanted	Passive
DOAS	Absorption, optical, fixed wavelengths	Path-averaging	Fixed, horizontal	Active, bistatic
Scintillometry	Observation of refractive index fluctuations	Path-averaging	Fixed, horizontal	Active, bistatic
Radiometry	Electromagnetic, fixed wavelength(s)	Averaging, profiling	Fixed, scanning, slanted, vertical	Passive
Tomography	Travel time, acoustic, fixed wavelength	Horizontal distribution	Fixed, horizontal	Active, multiple emitters and receivers

for an overview). If both scattering and absorbing processes are present, inversion algorithms have to be used to differentiate between the two processes.

If emission and detection takes place at the same site, the technique is called monostatic; if the detection is made at a different site than the emission, the technique is called bistatic. Monostatic remote sensing is relying on radiation that is scattered at an angle of 180°, bistatic techniques are used if other scattering angles are possible and maybe more favourable than the 180° angle.

Active remote sensing with pulsed signals usually allows for a range determination based on the travel time of the emitted signal.

Passive remote sensing is based on the detection of naturally emitted radiation from the atmosphere or radiation from an external radiation source (often from the sun), which has travelled through the atmosphere. The range determination usually requires an inversion algorithm.

3.2 RADAR

RADAR (radio detection and ranging) is an active remote-sensing technique that analyses the Rayleigh backscatter of electromagnetic waves at hydrometeors (Wexler and Swingle 1946). Occasionally, swarms of insects and birds lead to considerable echoes, too. Bragg backscatter from clear air is not usable unless strong moisture gradients – mainly in clouds are present (Knight and Miller 1998). Therefore, the RADAR technique has been developed to measure the location and intensity of precipitation (weather RADAR). If the Doppler shift of the backscattered radiation is analysed, it can also be used for horizontal wind measurements (Doppler-RADAR). RADAR is not a specific boundary layer observation tool. But it can give valuable information and it is presented here because it is the oldest ground-based remote-sensing technique. Recently, a special X band RADAR for boundary layer research has been constructed (Al-Sakka et al. 2009).

Typically, eight wavelength bands are differentiated in RADAR meteorology (see Table 3.2). L and P band and VHF RADAR devices are called windprofiler (see Section 3.2.1). The K band is sometimes divided into a Ka and Ku band.

A RADAR consists of a parabolic antenna with an emitter and receiver in its focal point. The antenna serves for the beam emission as well as for the signal detection. Because the aperture of the antenna and the RADAR wavelength are nearly of equal size, beam refraction is important and the RADAR beam cannot be focussed perfectly. During the emission side lobes are produced, leading to backscatter from the ground (ground clutter) and from obstacles away from the focal line (fixed echoes), which can disturb the measurements considerably. Known fixed echoes can be eliminated a posteriori. Strong electromagnetic radiation from a RADAR is dangerous to animals and human beings because it heats up the cells and can damage them.

The ratio between emitted and received power is described by the RADAR equation

$$P_{\mathrm{R}} = r^2 \frac{c\tau A\varepsilon}{2} P_0 \kappa^2 Z e^{-2\sigma r} + P_{\mathrm{bg}}, \tag{3.1}$$

Table 3.2 RADAR wavelengths and frequency bands following the International Telecommunications Union (ITU) Radar Band Nomenclature (see http://www.radioing.com/eengineer/bands.html for alternative definitions)

Frequency	Wavelength	Frequency bands	Application
20–300 MHz	1–15 m	VHF	Windprofiler
400–900 MHz	30–70 cm	UHF (P-Band)	Windprofiler
1–2 GHz	15–30 cm	L-Band/UHF	Boundary layer windprofiler
2–4 GHz	7–15 cm	S-Band/UHF	Precipitation
4–8 GHz	4–7 cm	C-Band	Precipitation
8–16 GHz	2–4 cm	X-Band	Precipitation
16–20 GHz	1–2 cm	Ku-Band	Precipitation
35 GHz	8.5 mm	Ka-Band	Precipitation, clouds
75–110 GHz	3 mm	W-Band	Clouds

here, P_0 denotes the emitted power (usually some hundreds of kW), P_R the received power (minimum detectable power is about 10^{-14} W) and P_{bg} the power of the background noise. ϵ is the efficiency of the antenna, A the effective area of the antenna, σ the scattering and absorption by particles soaring in the air, r the distance to the RADAR, τ the pulse duration (typically half a micro second), $\kappa^2 Z$ the backscatter cross-section ($\kappa^2_{\text{water}} = 0,93$, $\kappa^2_{\text{ice}} = 0,17$), and c the speed of light. The ratio of the two terms on the right-hand side of the RADAR equation is called signal-to-noise ratio (SNR). For the derivation of this equation Rayleigh scattering has been assumed. Therefore, the wavelength of the RADAR λ must be much larger than the diameter d of the precipitation particles ($\lambda > 20\, d$ for water droplets and $\lambda > 10\, d$ for snow flakes). This equation gives an integral value over all precipitation particles in the scattering volume. The range resolution Δr is given by

$$\Delta r = 0.5 c \tau. \tag{3.2}$$

With pulse durations of half a micro second, this gives a range resolution of 75 m. Because the received backscattered power is also proportional to the pulse duration, a trade off between maximum range and range resolution has to be made.

Usually, RADAR beams are emitted nearly horizontally or at a small angle to the horizontal plane. Therefore, the refraction of the beams is very important when computing their path length. Refraction can only be neglected for vertical beams. The magnitude of the refraction depends on the vertical gradient of the refraction index n of the atmosphere for electromagnetic radiation. This refraction index is mainly a function of the temperature and moisture distribution. In order to get convenient expressions, often a refractivity N is defined by

$$N = 10^6 (n - 1). \tag{3.3}$$

For this refractivity, the temperature and moisture dependence can be expressed as

$$N = 77.6\frac{p}{T} + 3.73 \cdot 10^5 \frac{e}{T^2}, \quad (3.4)$$

with the air temperature T in K, the air pressure p and the water vapour pressure e in hPa (Bean and Dutton 1968). The interesting problem for the refraction of the RADAR beam near the surface is whether the curvature of the beam is smaller or larger than the curvature of the Earth's surface. For this purpose a modified refractivity M is defined:

$$M(z) = 10^6 \frac{z}{R} + N = 157z + N, \quad (3.5)$$

with the height above the Earth's surface z and the Earth's radius R, both given in km. If $M(z)$ turns out to be constant with height z, then the RADAR beam follows the curvature of the Earth's surface always having a constant height above ground. If $M(z)$ decreases with height, the RADAR beam is bended towards the ground. In layers with a negative vertical gradient of M, RADAR beams are captured like in a wave duct. Such ducts can occur underneath of strong inversions. Vertical gradients of M between 0 and 78 km^{-1} are called superrefraction. Such conditions appear with stabile thermal stratification and a strong vertical decrease of moisture, and they lead to enhanced maximum ranges of a RADAR because the height above ground of the RADAR is only increasing slowly. Gradients of M between 78 and 157 km^{-1} are defined as normal propagation conditions, and gradients above 157 km^{-1} as subrefraction. Under subrefraction conditions, the RADAR is bended upward more than normally and the maximum range of the instrument is reduced because the beam is too high above ground in larger distances from the instrument.

A RADAR is usually operated in scanning modes. Rotation around the vertical axis while emitting the beam at a constant low elevation angle leads to circular maps of the precipitation distribution with the position of the RADAR in the centre of these maps. This mode is known as PPI (plane position indicator). Because the RADAR beam is usually bended less than the curvature of the Earth's surface, the height of the beam above ground increases with growing distance from the centre in a PPI mode. An elevation angle of 0.5° leads to height of the beam above ground of 4 km at a distance of 200 km. Successive scans with varying elevation angles can be used to compute circular maps, which artificially show the backscatter echoes from a fixed height above ground (CAPPI, constant altitude plane position indicator). A second scanning mode with constant azimuth and periodically varying elevation angle is called RHI (range height indicator) and serves to obtain vertical cross sections through precipitation areas (e.g. rain clouds and fronts). The maximum range of a RADAR is about 200 km.

The RADAR reflectivity Z in the RADAR equation depends on the size of the drop radii D_i in the measuring volume ΔV,

$$Z = \frac{1}{\Delta V} \sum D_i^6. \quad (3.6)$$

If the drop size spectrum $N(D)$ is known, this sum can be written as an integral

$$Z = \int N(D)D^6 \mathrm{d}D. \tag{3.7}$$

Marshall and Palmer (1948) for the first time formulated the distribution $N(D)$ as function of drop size D in mm and precipitation intensity R in mm h^{-1}:

$$N(D) = N_0 e^{-\Lambda D}, \tag{3.8}$$

with $N_0 = 8 \cdot 10^3$ m^{-3} mm^{-1} and $\Lambda = 4.1$ $R^{-0.21}$ mm^{-1} the well-known Z-R relation is found:

$$Z = 6! N_0 4.1^{-7} R^{1.47} = aR^b. \tag{3.9}$$

Marshall and Palmer (1948) found $a = 296$ and $b = 1.47$. Today, different a and b are used for different types of precipitation. Inserting this Z-R relation into the RADAR equation allows calculating the rain rate from the backscattered power P_R. Thus, principally, a quantitative rain measurement is possible. Because the reflectivity Z can take a huge range of values a logarithmic measure with $Z_0 = 1$ mm^6 m^{-3} is formed,

$$\mathrm{dBZ} = 10 \cdot \log_{10}(Z/Z_0). \tag{3.10}$$

The difficulty is to find the right Z-R relation. From the combination of drop size measurements and RADAR observations, the following typical values have been found: $a = 200$ and $b = 1.6$ for stratiform clouds and $a = 350$ and $b = 1.4$ for convective clouds. For snow $a = 2000$ and $b = 2$ is a possible choice. Table 3.3 gives typical values for the relation between the reflectivity and the rain rate using $a = 269$ and $b = 1.47$.

RADARs can be operated as pulsed RADARs or as frequency-modulated continuous-wave RADARs (FMCW-RADAR). The operational principle of S-band FMCW-RADARs, which are especially suited for boundary layer research, has recently been reviewed by Ince et al. (2003). While pulsed RADARs determine the range via the delay between the emission and the receiving of the pulse, continuous-wave RADARs principally cannot do so. Therefore, range-resolving continuous-wave RADARs must be operated with a constantly changing emission

Table 3.3 Typical RADAR reflectivities as function of the rain rate

Z [mm^6 m^{-3}]	DBZ	Rain rate [mm h^{-1}]	Notes
1	0	0.04	One drop with $D = 1$ mm m^{-3}
200	23	1.00	Light rain
15 625	42	15	One drop with $D = 5$ mm m^{-3}
257 800	54	100	Heavy rain
10^6	60	205	Very heavy rain (probably with hail)
10^7	70		Very large hail stones

frequency, i.e. as FMCW-RADARs. Usually, triangle-shaped frequency modes are used. The frequency modulation must lead to larger frequency differences between emitted and received signal than the expected Doppler shift due to motion of the scatterers along the line of sight. The advantages of FMCW-RADARs are continuous observation and less emitted power, which reduces the necessity for shielding.

3.2.1 Windprofiler

Windprofilers are Doppler RADAR instruments for the detection of the vertical wind profile working at frequencies of 50 MHz to 1 GHz (VDI 3786 part 17, 2007). The respective wavelengths are between 6 and 0.3 m. Windprofiler operating at up to about 300 MHz is called VHF windprofiler (very high frequency), those operating at higher frequencies as UHF windprofiler (ultra high frequency) (see Table 3.2). In the VHF frequency range we observe usable Bragg backscatter at temperature and especially at moisture fluctuations in the atmosphere. Water and rain drops do not absorb in this range; therefore, windprofiler measurements are not disturbed by the presence of clouds and rain. VHF Windprofiler give profiles from the stratosphere and mesosphere, but have insufficient lowest range gates of about 1 km. UHF windprofiler have a lower range and are used for the observation of the troposphere and can be used for boundary layer studies as well. The disadvantage of operating a windprofiler at such high frequencies such as 915 MHz is the enhanced sensitivity to hydrometeors. This hampers the measurement of vertical velocity in case of rain. The measurement of horizontal velocity is still possible, if it is assumed that the rain drops are advected with the mean wind. On the other hand, the vertical range is enhanced in case of precipitation compared to clear air profiling (Ecklund et al. 1988). Mainly at lower levels, ground clutter contaminates estimations of the vertical velocity component and thus a quasi-systematic bias of the order of -10 cms^{-1} exists (Caccia et al. 2004).

Due to the high propagation speed of electromagnetic waves (in the order of 3×10^8 m s^{-1}), the height resolution of windprofilers is only about 100 m. The range depends on the chosen frequency (from about 1 km at 1 GHz to about 70 km at 50 MHz) and the emitted power. The size of the antenna is proportional to the wavelength. For 6 m wavelength an area of about 100 m times 100 m is required, for shorter wave lengths 10 m by 10 m with yagi elements or even a bowl-shaped antenna can be sufficient (Strangeways 2003).

At Lindenberg observatory of the German Weather Service, a 1290 MHz boundary windprofiler is operated (Fig. 3.2). It has a range of 4–6 km and has been extended to a RASS (see below) by the addition of an acoustic emitter.

Windprofilers work with the Doppler-beam-swinging (DBS) technology for the detection of profiles of the three-dimensional wind vector. Here, three to five beams are emitted one after the other into three to five different directions. One of the three or five directions is always the vertical direction, the other directions are tilted by 15–20° from the vertical. The azimuthal directions of the tilted beams usually

Fig. 3.2 Boundary layer wind profiler of the German Weather Service at the Richard Aßmann observatory in Lindenberg. The electromagnetic antenna is to the *left*, the acoustic sound source to the *right*

differ by 90°. The half-power beam width is of the order of 8° (Caccia et al. 2004). Like RADAR instruments the emitted beams from windprofiles have side lobes, which can lead to the recording of fixed echoes. Wind speed measurements from these instruments are disturbed by moving obstacles (swaying trees, wind energy converters). Shielding to the sides can reduce the disturbance by fixed echoes.

3.2.2 Micro Rain RADAR

Micro rain RADARs are also based on the RADAR technology. They are frequency-modulated continuous-wave (FM-CW) RADAR instruments, which emit a vertically directed beam with a wavelength of 1.25 cm (24.1 GHz). Micro rain RADARs record not only the backscattered intensity (reflectivity) but also analyse via the Doppler shift of the backscattered signal the falling velocity of the scattering precipitation particles. With this coinciding information on fall speed and drop size, a quantitative precipitation measurement is possible (Peters et al. 2002, 2005). The instrument has an emitting power of 50 mW and can resolve 28 height range intervals between 35 and 200 m. If a height interval of 100 m is chosen, a layer from 300 m to 3000 m above ground can be analysed. The minimum time resolution is 10 s; typically 60 s is used. The small instrument which weighs just 12 kg operates with a parabolic antenna of 60 cm diameter that emits the 2° wide beam. The device is usually mounted on a mast of a few metres height and does not require a radome (Fig. 3.3).

Fig. 3.3 Micro Rain RADAR (courtesy of METEK GmbH, Germany)

3.3 SODAR

A SODAR is an active acoustic remote-sensing device that (nearly) vertically emits sound pulses and receives and analyzes the backscattered part reflected from refraction index fluctuations in the atmosphere. The name of this device is formed in analogy to RADAR and means sound detection and ranging (Gilman et al. 1946). An overview of the basic principle of sounding with a SODAR and the history of the development of this instrument is described in Peters (1991). SODARs analysing the Doppler shift of the backscattered sound pulses for the derivation of the vertical wind profile are called Doppler-SODARs.

Sound waves are scattered at turbulent density fluctuations in the atmosphere, which are supposed to move with the mean wind, because the refraction index for sound waves is changing at the boundaries of these fluctuations. For a monostatic SODAR (the receiver is identical with the emitter), the intensity of the backscattered signal depends only on these density fluctuations; for a bistatic SODAR (the receiver is deployed away from the emitter) also velocity fluctuations in the atmosphere contribute to the intensity of the backscattered signal. The density fluctuations are mainly determined by temperature fluctuations. From a non-turbulent quiet atmosphere at rest without any temperature gradients, no backscattered signal can be received with a SODAR. For optimal backscatter, the spatial size of the temperature gradients in the atmosphere should be about half the acoustic wavelength (Bragg condition). This condition can be fulfilled by turbulent fluctuations as well as by temperature inversions. When deducing information on the atmospheric state from the backscattered signal, one has to differentiate between these two possibilities. This can be done either from an assessment of the general weather conditions or (if the SODAR is a Doppler-SODAR) from an analysis of the variance of the simultaneously recorded vertical velocity. High variances indicate thermal forcing, and the backscatter intensity is proportional to the turbulence intensity; low variances indicate stable layering and the backscatter intensity should be proportional to mean vertical temperature gradients (i.e. inversions).

Depending on the technology how the sound beams are focussed, a difference is made between "classical" SODARs, which had been developed first, and phased-array SODARs. Classical SODARs use three large tiltable antennas to focus the beams (Fig. 3.4). Several large sound transducers are fixed at the bottom of each of these antennas, and within one antenna they are operated simultaneously. The sound transducers serve as emitters as well as receivers. Phased-array SODARs have a larger number of smaller sound transducers (often 32 or 64), which are regularly arranged on a quadratic plate with an area of about 1 m^2. By using the interference principle (enhancement and extinction), slanted and vertical beams are formed by operating the various sound transducers with specified time delays among each other (Fig. 3.5). While with classical SODARs, the antennas simultaneously serve as a shield to protect the environment from the sound pulses as well as the SODAR from disturbing ambient noise, phased-array SODARs need a shield (baffles) around them. Nevertheless, even if optimal lateral shields are erected, the sound pulse is still

Fig. 3.4 Three-antenna sodar for wind and turbulence profile measurements up to about 1000 m above ground

Fig. 3.5 Phased-array sodar for wind and turbulence profile measurements up to several hundreds of meter above ground. *Left*: outside view showing the cuffs protecting the instruments from ambient noise. The ragged top of the cuffs is made in order to reduce the formation of side lobes. *Right*: View to the inside of the cuffs showing part of the sound transducer array on the bottom

detectable for human ears at a distance of several hundreds of metres. Thus, siting of SODARs close to residence areas and office buildings has to be avoided or need the explicit approval of the people living and working there.

Doppler-SODAR instruments for the measurement of wind profiles work with the Doppler-beam-swinging (DBS) technology similar to windprofilers. Here, three to five beams are emitted one after the other into three to five different directions. One of the three or five directions is always the vertical direction, the other directions are tilted by 15–20° from the vertical. The azimuthal directions of the tilted beams usually differ by 90°. Because the following beam cannot be emitted before the backscatter of the previously emitted beam has been received in order to avoid disturbances, one measurement cycle of a SODAR lasts six to ten times the time a sound pulse needs to travel up to the maximum vertical range of the instrument. This maximum vertical range depends on the emitted power and the used sound frequency and ranges from about 200 m at 4500 Hz to about 1 km at 1500 Hz. The frequency dependence is due to the different absorption of sound waves of different frequency in the atmosphere. The higher the frequency is the stronger is the absorption. And 4500 Hz-SODARs with a maximum range of up to 200 m are frequently called MiniSODARs. For a three antenna SODAR operating at 1500 Hz, this means a length of a measurement cycle of about 20 s. The measurement of the wind speed is based on the analysis of the Doppler shift between the emitted and the received signal. The height of the reflecting air volume is computed from the travel time of the sound signal. The minimum range gate is around 20–30 m above the instrument, the vertical resolution depends on the pulse duration and is in the order of 5–20 m.

Using an emitting power of about 1 kW leads to a backscattered signal with a power of just of 10^{-15} W. This is below the hearing threshold of the human ear

(10^{-12} W) but well above the noise produced by the Browian movement of the atmospheric molecules (6×10^{-19} W). The used sound frequencies of 1500–4500 Hz are well within the sensitivity range of the human ear.

The ratio between emitted and received power is described by the SODAR equation,

$$P_R = r^{-2}(c_s \tau A \varepsilon/2) P_0 \, \beta_s \, e^{-2\sigma r} + P_{bg}, \tag{3.11}$$

with the received power P_R, the emitted power P_0, the antenna efficiency ϵ, the effective antenna area A, the sound absorption in air σ, the distance between the scattering volume and the instrument r, the pulse duration τ (typically between 20 and 100 ms), the backscattering cross-section β_s (typically in the order of 10^{-11} m^{-1} sr^{-1}), the sound speed c_s, and the background noise P_{bg}. The background noise also comprises contributions from ambient noise having the same sound frequency, e.g. traffic noise. The ratio of the two terms on the right-hand side of the SODAR equation is called signal-to-noise ratio (usually abbreviated as SNR). The backscattering cross-section β_s is a function of the temperature structure function $C_T{}^2$ (Tatarskii 1971). For a monostatic SODAR, we find (Reitebuch 1999) when using the wave number $k = 2\pi/\lambda$,

$$\beta_s(180°) = 0.00408 \, k^{1/3} \, C_T^2/T^2. \tag{3.12}$$

This equation forms an average over all scattering elements within the atmospheric volume that is hit by the cone-shaped beam. The pulse duration τ determines the height resolution of the instrument via the relation

$$\Delta z = 0.5 \, c \, \tau. \tag{3.13}$$

The SODAR equation above showed that the backscattered power was proportional to the pulse duration, too. Therefore, the choice of the pulse duration is a trade-off between height resolution (preferably short pulse durations) and maximum range (preferably long durations).

The backscattered signal of a SODAR is representative for an atmospheric volume. For pulse durations of 100 ms we simultaneously receive backscatter from a volume of 33 m depth. In a height of 500 m above ground this volume has a radius of 44 m (assuming an opening angle of the emitted sound beam of 5°). Additionally, the three-dimensional wind information has to be integrated from a full measurement cycle over one vertical and two to four tilted beams. This increases the radius of the detected atmospheric volume in 500 m height to about 150–200 m. Further, in order to enhance the signal-to-noise ratio, averages over all measurement cycles with 10–30 min are formed. This makes the SODAR a volume and time averaging measurement device.

In addition to mono-frequency SODARs, multi-frequency SODARs are also available, which emit several subsequent pulses with different frequencies within one shot. Field tests have proved significant advantages of the multi-frequency technique. The use of, e.g., eight different frequencies halves the minimal acceptable signal-to-noise ratio compared to single-frequency sounding. Moreover, the multi-frequency mode improves the accuracy of instantaneous values of measured parameters and significantly increases the reliability in recognizing noisy echo-signals. Further details on the principle of multi-frequency SODARs are given in Kouznetsov (2009).

Like RADAR beams SODAR beams also have side lobes. Therefore, fixed echoes can severely influence and disturb SODAR measurements. This requires a good site selection for these instruments. SODARs have to be sited away from obstacles like buildings, trees, and electric cables.

3.3.1 SNODAR

A special SODAR development for extreme cold environments is a SNODAR (Surface layer NOn-Doppler Acoustic Radar), which is designed to measure the height and turbulence intensity of the atmospheric boundary layer on the Antarctic plateau. This is for example of relevance for astronomers wishing to plan future optical telescopes there. SNODAR works by sending an intense acoustic pulse into the atmosphere and listening for backscatter off inhomogeneities resulting from temperature gradients and wind shear. The theory of operation is very similar to that of the well-known underwater sounding techniques of SONAR. SNODARs are monostatic acoustic radars with a minimum sampling height of 5 m, a range of at least 200 m, and a vertical resolution of 1 m. SNODARs operate at frequencies between 4 and 15 kHz. Such high frequencies propagate relatively well in the low temperature of the Antarctic atmosphere (Bonner et al. 2008).

3.4 RASS

A radio-acoustic sounding system (RASS) operates acoustic and electromagnetic sounding simultaneously (Marshall et al. 1972). This instrument is able to detect acoustic shock fronts of the acoustic pulses and to determine their propagation speed from the Doppler shift of the backscattered electromagnetic waves. This propagation speed is equal to the speed of sound, which in turn is a known function of air temperature and humidity. Two different types of RASS have been realised (Engelbart and Bange 2002): a Bragg-RASS and a Doppler-RASS.

3.4.1 Bragg-(Windprofiler) RASS

A Bragg-RASS is a windprofiler (see Section 3.1.1 above) with an additional acoustic emitter. When the Bragg condition is fulfilled (Fig. 3.6), i.e. the wavelength of the sound waves λ_a is half the one of the electromagnetic waves λ_e, then there is

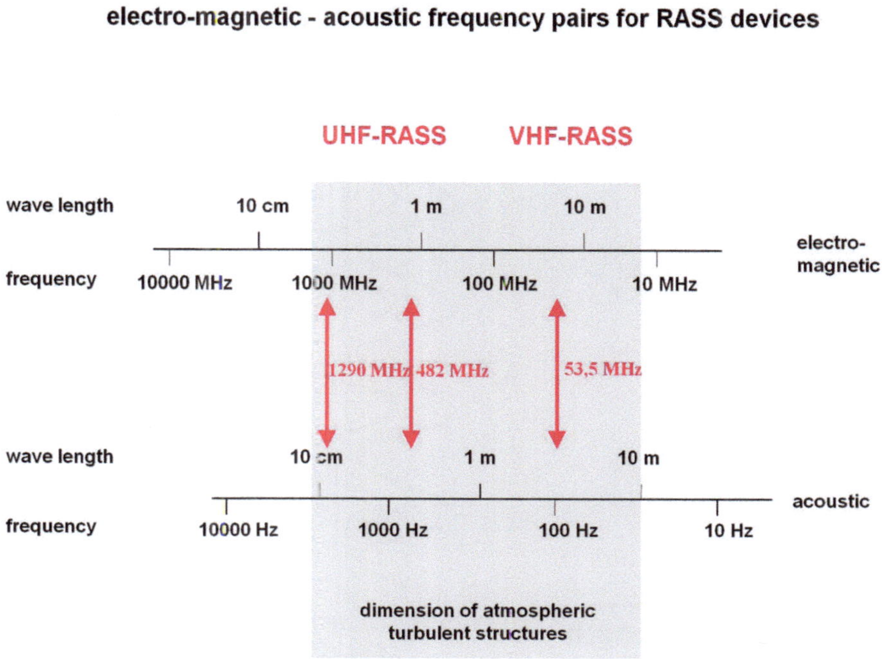

Fig. 3.6 Cut from Fig. 3.1 showing the frequency pairs used for different RASS combinations. The two combinations to the left are utilized for boundary layer profiling. The acoustic wave lengths are shifted according to the Bragg condition (see also Fig. 3.7)

optimal backscatter of the electromagnetic waves from the acoustic waves (Fig. 3.7). The electromagnetic signal is emitted at a fixed frequency, but the emitted sound signal is a chirp signal with varying frequency f_a. From the sound wave length $\lambda_{a,B}$ at which optimal backscatter occurs the propagation speed of the sound signal can be determined via the following dispersion relation:

$$c_a = \lambda_{a,B}/2 * f_a. \tag{3.14}$$

For a VHF windprofiler operating at 50 MHz a sound frequency of about 100 Hz is used, for a UHF windprofiler operating at 1 GHz a sound frequency around 2 kHz is most suitable to fulfil the Bragg condition. Because the attenuation of sound waves in the atmosphere is strongly frequency-dependent, a UHF RASS can detect temperature profiles up to about 1.5 km height whereas a VHF RASS can observe temperature profiles throughout the troposphere.

3.4.2 Doppler-(SODAR) RASS

A Doppler-RASS (or SODAR-RASS) is a SODAR with an additional electromagnetic emitter and receiver (Fig. 3.8) operating at a frequency $f_{e,0}$. From the Doppler

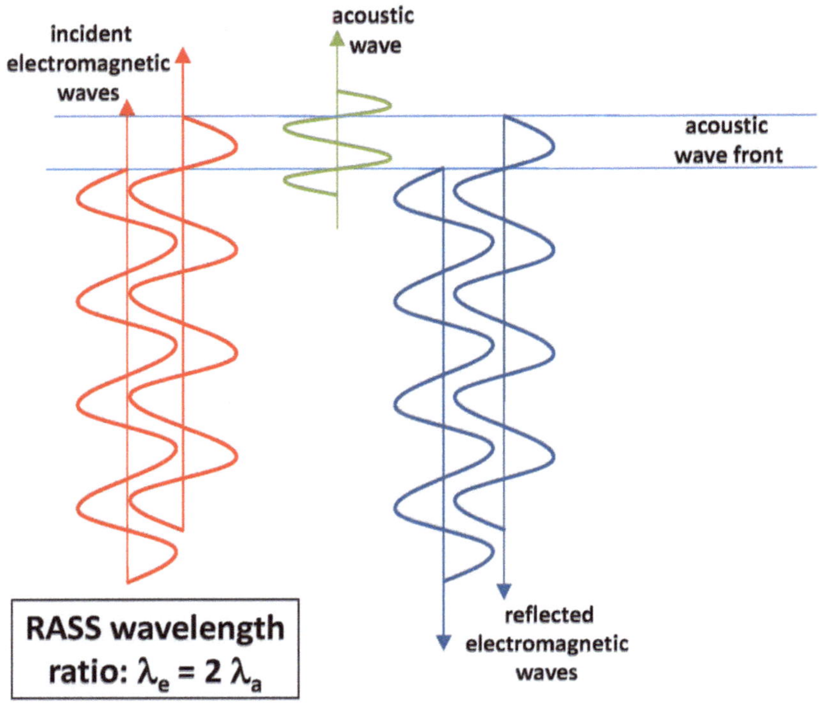

Fig. 3.7 Schematic explanation of the Bragg condition for RASS devices. *Green*: acoustic pulse (shock wave front) travelling upward

shift Δf_e of the electromagnetic radiation which is backscattered at the density fluctuations caused by the sound waves, the propagation speed c_a of the sound waves is determined,

$$c_a = -0.5\ c * \Delta f_e / f_{e,0}), \tag{3.15}$$

where c denotes the speed of light. A Doppler-RASS like a Bragg-RASS also emits a chirp sound signal in order to assure that the Bragg condition is optimally met due to the varying temperature over the entire height range.

The so determined propagation speed c_a is a sum of the speed of sound c_s and of the vertical movement of the air w within which the sound waves propagate,

$$c_a = c_s + w. \tag{3.16}$$

The vertical air speed component w can be determined separately from the Doppler shift of the backscattered electromagnetic clear-air signal when operating a Bragg-RASS or from the Doppler shift of the backscattered acoustic signal when operating a Doppler-RASS. Using the definition of the acoustic temperature, the height profile of c_s can then be converted in a height profile of the acoustic temperature T_a. For many purposes this acoustic temperature is a

Fig. 3.8 Radio-acoustic sounding system (Sodar-RASS) for profile measurements of wind, turbulence, and temperature. The acoustic sounder (sodar) with its three white antennas is in the middle, the emitting radio antenna is to the left and the receiving radio antenna is to the right

sufficiently accurate approximation of the virtual air temperature. The maximum range of a SODAR-RASS is usually less than thousand metres so that such an instrument covers the lower part of the boundary layer. Due the high vertical resolution and the low minimum range of about 30–40 m, which is comparable to the abilities of a SODAR, a SODAR-RASS is well suited for the detection of shallow nocturnal boundary layers.

3.5 LIDAR

A light beam emitted into the atmosphere can interfere with the air in different ways. Alterations may occur due to elastic scattering (Rayleigh scattering at atoms and molecules, Mie scattering at aerosols and water droplets (especially at clouds and rain drops)), inelastic scattering (Raman scattering), fluorescence, absorption, and Doppler shift. Based on these different interactions, five types of LIDARs (Light detection and ranging) have been developed to actively and remotely analyse trace substance concentrations in the atmosphere (Weitkamp 2005). As fluorescence LIDARs do not have any applicability in the boundary layer, four types are of interest here:

- Backscatter LIDARs and ceilometers (elastic scattering)
- differential absorption LIDAR (DIAL)
- Raman LIDAR (inelastic scattering)
- Doppler wind LIDAR

In contrast to sound waves and electromagnetic waves used in RADARs and windprofilers, LIDAR wavelengths are several orders of magnitude shorter than the geometric aperture of the emitting optics, therefore diffraction is negligible and side lobes to the main beam do not appear. Thus, fixed echoes from obstacles away from the optical axis of a LIDAR do not influence the measurements as they do in RADAR and SODAR measurements. A LIDAR detects the backscattered light with a telescope mounted in parallel to the emitting optical axis. The lateral distance between the two optical axes is a few tens of centimetres; therefore, the detection of near-field backscattering (up to a few hundred metres from the instrument) is not possible. Partly, this effect is desired because the intensity of the backscattered light from such short distances would be by far too high and would blind the receiving photomultipliers. Range determination is done from the travel time of the pulsed signal. A few LIDAR systems work with beam focussing (see Section 3.5.4.2 below) or use coherent tomography principles (Section 3.5.4.3 below) for the range determination.

Until recently, optical methods had the disadvantage of a quite low range (and hence also height) resolution of about 50 m and large lowest range gates of several hundreds of metres. For example, a commercially available solid-state system operates at a minimum range of ~400 m with range gates that could be as low as 50–75 m (Hannon 2004). The range resolution is limited by the pulse duration, which cannot be minimised unboundedly since there is a direct trade off against the velocity resolution. The lowest range gate is caused by the unavoidable switching between transmitting and receiving. A larger lowest range gate also helps to safeguard the instrument against blinding by extremely high backscatter intensities from nearby reflecting objects. Acquisition of VAD conical scan data at low elevation angles leads to a partial compensation for the poor height resolution, but this in turn is achieved at the expense of an even larger measurement volume. These drawbacks had until recently limited the use of optical remote-sensing techniques for wind speed mainly to the determination of wind shear at airports (see also VDI 3786-14, Hannon 2004, and Chan et al. 2006). A comparison between the mentioned commercially available system and a RADAR wind profiler can be found in Vogt and Wieser (2006). In the meantime, enhancements in laser technology, in electronic signal processing, and innovative means of range determination for backscattered signals have decreased these problems. As a consequence, e.g. optical wind measurement techniques have recently become more suited to the requirements of wind energy applications. Furthermore, the usage of infrared light makes this technology eye-safe, a mandatory requirement for an automatic measurement technique running unattended for longer time periods.

Optical methods have no negative influence on the environment. They do not produce noise or electromagnetic stray fields and they are not affected by these disturbances. Eye safety has to be ensured when operating LIDARs with strong light sources (laser) following international guidelines such as IEC 60825-1.

3.5.1 Backscatter (Aerosol) LIDAR

A backscatter LIDAR is employed to determine aerosol and trace gas profiles in the atmosphere. The first such application is described in Fiocco and Smullin (1963). The intensity of the backscattered signal P_R is expressed by the LIDAR equation (Menut et al. 1999):

$$P_R(\lambda, r) = r^{-2}(c\tau A\varepsilon/2)\, P_0\, [\beta_m(\lambda, r) + \beta_p(\lambda, r)]\, e^{-2\sigma r} + P_{bg}, \qquad (3.17)$$

with the distance r between the LIDAR and the backscattering object, the speed of light c, the pulse duration τ, the antenna area A, a correction term for the detector efficiency and losses due to the lenses ϵ, the emitted energy P_0, the backscatter coefficient for molecules β_m and for particles β_p, the absorption of light in the atmosphere σ, and the background noise P_{bg}. The background noise also comprises scattering of light from other sources (e.g. the sun) into the optical axis of the receiving telescope. The ratio of both terms on the right-hand side of (3.17) is the signal-to-noise ratio (often abbreviated as SNR). The noise- and distance-corrected signal P_{RSC} can be calculated from P_R:

$$P_{RSC}(\lambda, r) = (P_R - P_{bg})\, r^2 = (c\tau A\varepsilon/2)\, P_0\, [\beta_m(\lambda, r) + \beta_p(\lambda, r)]\, T^2(\lambda, r), \quad (3.18)$$

with the transmission $T = e^{-\sigma r}$. Normalizing with the emitting power P_0 yields

$$P_{RC}(\lambda, r) = ((P_R - P_{bg})/P_0)\, r^2 = (c\tau A\varepsilon/2)\, [\beta_m(\lambda, r) + \beta_p(\lambda, r)]\, T^2(\lambda, r). \quad (3.19)$$

Due to the varying aerosol content of the atmosphere the measurement of a trace gas concentration with a simple backscatter LIDAR is not possible, because a separation of the two influences of β_m and β_p on P_R is not possible without additional assumptions. Furthermore, also assumptions on the transmission T are necessary, the most fundamental problem of LIDAR trace substance measurements. Usually, this problem is circumvented by assuming an aerosol type-dependent fixed ratio between backscatter and absorption in spectral ranges free of molecule absorption.

3.5.1.1 Ceilometer

Ceilometers are simple backscatter LIDARs that merely record the optical backscatter intensity in the near infrared at about 0.9 μm (Fig. 3.1). The light emission into the vertical direction is pulsed (10 kHz) with a pulse duration of 110 ns. Due to the chosen wave length and the restricted emission power (11 W peak), these instruments are eye-safe (Münkel et al. 2003, Weitkamp 2005) and can be run mostly unattended. Originally, ceilometers have been developed for the measurement of

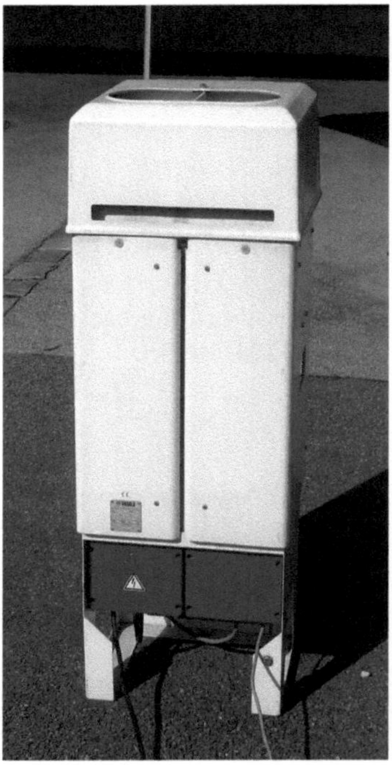

Fig. 3.9 Ceilometer

the height of the cloud base (ceiling). Today, they can frequently be found at airports and in their surroundings. Figure 3.9 shows a ceilometer, which is about a metre high. The emitted and received signal pass through the window on the top side of the instrument.

A closer investigation of the lower dynamic range of the backscatter intensity in the recent years has proven that the backscatter profiles carry useful information on the vertical aerosol distribution in the atmosphere up to a height of several kilometres above the instrument. Range determination for the profiles is done via the signal travel time. Two different types of ceilometers are in use: those with two optical axes (one for the emitted light beam and one for the receiving telescope) and those with one optical axis where the emitted beam is sent through a small hole in the mirror of the receiving telescope. Instruments with one optical axis are able to yield useful information already from a distance of about 30 m. Instruments with two optical axes have an insufficient overlap in the lower 150 m so that backscatter information from this height range cannot reliably be interpreted (Münkel et al. 2007).

3.5.2 Differential Absorption LIDAR

The basic idea of differential absorption LIDARs (DIALs) is the emission of light pulses at two neighbouring frequencies (Weitkamp 2005). The frequencies are chosen such that one frequency is optimally absorbed by the sought trace gas (λ_{max}) while the other frequency (λ_{min}) experiences nearly not absorption. The two frequencies are chosen so close to each other that no difference in the aerosol backscatter is detectable and the influence of the aerosol can be eliminated by subtracting the two signals from each other. The detected intensity of the backscattered light at frequency λ_{max} is inversely proportional to the concentration of the absorbing trace gas. Principally, a DIAL is a backscatter LIDAR. The ratio of the backscattered optical intensities at the two frequencies is (Colls 2002)

$$I(\lambda_{max})/I(\lambda_{min}) = \exp(-2r/N\sigma), \tag{3.20}$$

where r is the distance of the backscattering air volume, N is the number concentration of the absorbing trace gas, and σ is the absorption cross section of the absorbing gas.

The following types of laser are available for a DIAL: (a) dye laser with frequency multiplier (the wave lengths may be chosen arbitrarily but these instruments require much attendance and adjustment, so that they are not suitable for monitoring purposes), b) Nd:YAG laser (1064 nm) with frequency multiplier (266 nm) and subsequent Raman shift in deuterium (289 nm) and/or hydrogen (299 nm), (c) krypton fluoride excimer laser (248 nm) and subsequent Raman shift in deuterium (268, 292, 319 nm) and/or hydrogen (277, 313 nm). Wulfmeyer (1998) developed an enhanced DIAL based on an alexandrite laser with reduced noise and reduced systematic errors that is able to measure water vapour fluctuations for moisture flux determinations.

A DIAL yields a vertical resolution of about 75 m for ozone measurements, a temporal resolution of about 60 s for a height range between several hundreds of metres and many kilometres.

3.5.3 Raman-LIDAR

A further possibility for the detection of trace gas profiles in the atmosphere is the operation of a Raman LIDAR, which records radiation from Raman scattering at trace gas molecules (Cooney 1970, 1972; Ansmann et al. 1990; Turner et al. 2002; Weitkamp 2005). With normal elastic scattering a molecule absorbs a photon, enters an excited state, and immediately emits (backscatters) the photon again while returning to the ground state. With inelastic Raman scattering, the molecule does not return to the ground state but to a neighbouring energy level so that a photon with a slightly changed (usually reduced) frequency is emitted. The observed frequency shift is characteristic for a given trace gas species (Fig. 3.10). The procedure is named after the Indian physicist, Chandrasekhara Venkata Raman, who first reported on the experimental discovery of this frequency shift in 1928. The disadvantage of the method is that the inelastically backscattered signal intensities are two to three orders of magnitude lower than with the elastic Rayleigh scattering. Therefore, measurements with a Raman LIDAR are possible only at nighttime or with special provisions to exclude other scattered light.

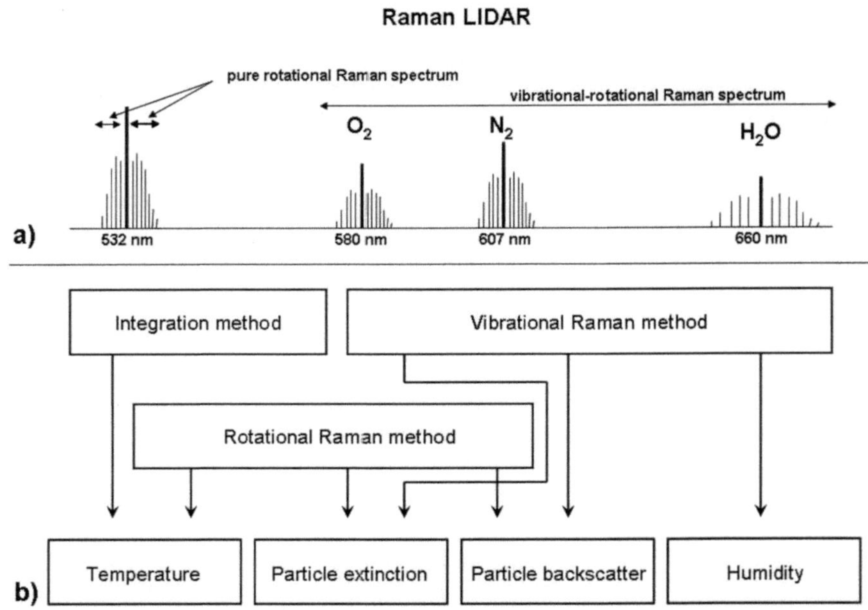

Fig. 3.10 (**a**) Schematic of the arrangement of Raman-shifted backscattered wavelengths, and (**b**) evaluation methods for a Raman-LIDAR

In a relatively cool atmosphere, the scattering air molecules are nearly always in their energetic ground state. Therefore, light from Raman scattering has usually less energy than the radiation emitted from the LIDAR (Stokes scattering, small lines to the right of the main lines in Fig. 3.10a). With rising temperature an increasingly number of air molecules can be in excited states, so that Raman scattering also from jumps to energy levels below the initial level becomes possible (Anti-Stokes scattering, small lines to the left of the main lines in Fig. 3.10a). Then, scattered light with a higher frequency than the emitted one can be detected as well. But this more energetic radiation from anti-Stokes scattering is weaker than the less energetic radiation from Stokes scattering.

A Raman LIDAR permits with different evaluation methods the detection of trace gas profiles from characteristic frequency shifts, the determination of type and size of aerosols from extinction measurements, the classification of aerosols from depolarization measurements, and the determination of air temperature from analysing the temperature-dependent ratio of different Raman frequencies (see Fig. 3.10b). While stratospheric temperature profiles, which can be derived with the integration method (Keckhut et al. 1990), are irrelevant for the boundary layer, temperature and particle profiles in the troposphere are deducible with the rotational Raman method (Cooney 1972; Behrendt and Reichardt 2000). Humidity and particle profiles are obtainable with the vibrational Raman method (Cooney 1970; Behrendt et al. 2002).

3.5.4 Doppler Wind LIDAR

In the following, currently available optical remote-sensing methods based on the analysis of the Doppler shift are presented. Doppler wind LIDARs are special backscatter LIDARs (see Section 3.4.1). In recent years special Doppler wind LIDARs for measurements in the near-surface layer from about 40 m up to about 200 m have been developed and are presently used for wind energy assessments (for a review see: Emeis et al. 2007). Different methods for range determination will be addressed.

3.5.4.1 Range Determination by Signal Delay

Range-resolved remote-sensing systems transmit signals in pulses, which are then scattered by atmospheric inhomogeneities or suspensions (e.g. aerosol, droplets), sending a small fraction of the transmitted energy back to the receiver. Distance to the measurement volume is determined from the time of flight of the signal pulse. Overviews of the state of the art of LIDAR techniques for wind and turbulence measurement using signal delay for the range determination have been given by Hardesty and Darby (2005) and Davies et al. (2003). Two common approaches to determining the frequency of the return signal for Doppler analysis are coherent (heterodyne) detection and direct (or "incoherent") detection, which will be described in this section.

Mie Scattering: Coherent Doppler LIDAR

Coherent or heterodyne detection of wind speeds using Doppler LIDAR relies on the ability to generate two signals having a precise offset in frequency. The signal transmitted to the atmosphere comes from a primary laser, variously called a "master oscillator" (MO), a "pulse laser" (Grund et al. 2001), or a "power oscillator" (PO; e.g. Post and Cupp 1990). Generally, the second or reference signal comes from a second, low-power, continuous-wave laser, or "local oscillator" (LO), although in some systems the frequency difference is produced by splitting the signal from the MO and offsetting the frequency of one of the two resulting signals. The wavelength of coherent LIDARs has typically been in an eye-safe region of the infrared spectrum, between 1.5 μm (solid-state lasing source) and 10.6 μm (CO_2 source), where atmospheric scattering is dominated by aerosol. In a coherent LIDAR receiver the backscattered return signal from the atmosphere is optically mixed with the coherent reference-frequency radiation from the LO at the system detector. The detector senses the beat frequency between the return signal and the reference signal (heterodyne detection), and the Doppler shift is calculated from this beat frequency. A detailed description of the configuration, design characteristics, trade offs, and data processing issues of a coherent Doppler LIDAR system is provided in Grund et al. (2001).

Fig. 3.11 Coherent Doppler wind LIDAR for wind energy assessment. The scanning infrared laser beam is sent (and the backscatter is received) through the circular opening on the top side of the instrument

Because of the sensitivity of coherent LIDAR systems, they can be made relatively compact, yet maintain useful performance standards, including maximum ranges (several kilometers to ~20 km), range resolution (less than 100 m), and velocity precision (a few 10's of cm s^{-1}). Such systems have a very narrow beam, which allows sampling close to the Earth's surface, including near topography and buildings. Because of these capabilities, they are well suited for atmospheric boundary layer applications (an example for such a device is shown in Fig. 3.11). A special scanning wind LIDAR for a height range between 40 and 200 m for wind energy assessments is available since a few years (Wächter et al. 2009).

Rayleigh Scattering: Incoherent Doppler LIDAR

An incoherent (or "direct detection") Doppler LIDAR (Canin et al. 1989) is an alternative for the measurement of atmospheric wind speeds in air masses with low or vanishing aerosol loads. These pulsed systems usually operate in the UV and rely on contributions from both aerosol and the enhanced molecular (Rayleigh) scattering at shorter wavelength to provide the return signal. A Fabry-Perot interferometer (Abreu et al. 1992) can be used to analyse the Doppler shift of the Rayleigh backscatter. While the capabilities of incoherent LIDAR are impressive (they are being considered for space-based global wind measurements), their current size, cost, and complexity suggest the technique is inappropriate for wind energy applications considered here (Harris et al. 2006).

3.5.4.2 Range Determination by Beam Focusing

A different method for range determination is beam focusing (Fig. 3.12). Range attribution is made by the assumption that the overall majority of the backscattered light is from the focal range. Such wind LIDARs (Fig. 3.13 shows a currently available beam-focusing Doppler-LIDAR for a height range of up to about 200 m) emit a continuous-wave (CW) beam, and interrogation of the wind speed at a given range is achieved by focusing, rather than by the time-of-flight method of pulsed systems. The interferometer of the system cannot distinguish between air motion towards and away from the LIDAR, and this leads to an ambiguity of 180° in the derived value of wind direction. This is easily resolved with reference to a simple wind direction measurement at a height of a few metres. Several different measuring heights between 5 and 150 m above ground can be programmed. The profile of the three-dimensional wind vector is yielded by scanning a cone with 30° half angle once per second (Banakh et al. 1995). Hence the diameter of the measurement volume is 173 m at 150 m height. The probe length increases roughly as the square of the height; as an example the vertical resolution is ~±10 m at 100 m height. Strong reflections from particles and other moving objects outside the focal range (e.g. due to smoke, fog or birds) can lead to spurious Doppler returns (Harris et al. 2001), but these effects can be recognised and mitigated by signal processing techniques. An intercomparison exercise between a beam-focussing wind LIDAR and cup anemometers at heights between 40 and 100 m above ground on the wind energy test range of the Danish research centre Risø in Høvsøre has yielded very good agreement (Smith et al. 2006). An application for wind energy assessment is described in Kindler et al. (2007).

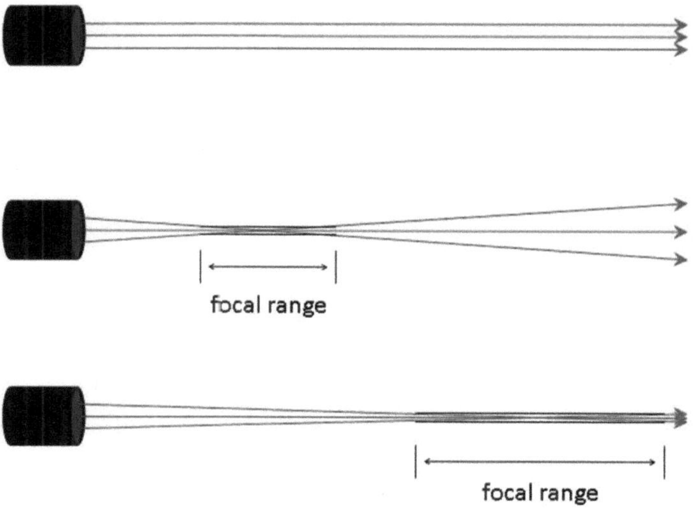

Fig. 3.12 Schematic of a beam-focussing Doppler wind LIDAR. Due to geometrical reasons the focal length is increasing with increasing distance from the instrument

Fig. 3.13 Beam-focussing Doppler wind LIDAR. The laser beam is sent and received through the slanted top of the grey upper part of the main instrument. The small mast to the left carries the wind direction sensor

The compactness of these new LIDAR systems also opens the potential to measure horizontally oncoming gusts upstream of a wind turbine at hub height. This capability was first successfully demonstrated in a proof-of-principle experiment in 2003, in which a beam-focussing LIDAR was mounted on the nacelle of a Nordex N90 turbine (Harris et al. 2006). Due to the high data rate of 10 Hz, the short interruptions of the beam by the blades of the turbine did not impede the measurement. A wind speed of 15 m s^{-1} and a focusing of the beam at 150 m distance permits a warning time of 10 seconds in order to adapt the turbine. With 25 m s^{-1} wind speed the warning time is 6 s, still considerably longer than the timescale for blade pitch adjustment. More research is needed to quantify the benefit of incorporating LIDAR as part of the turbine control system; such a scheme might involve staring directly upwind, but it could also employ a scan pattern in order to probe a wider area in front of the blades. A similar LIDAR system has also been installed in a rearward-looking configuration on a test turbine by Risø (Bingöl 2005); a programmable scanner permits examination of turbine wake behaviour in space and time.

3.5.4.3 Optical Coherence Tomography for Range Determination

In addition to signal delay and beam focusing, further attempts have been made to enhance the range resolution of optical remote sensing. One approach is to use optical coherence tomography, which has been applied quite successfully in medical examinations (Fercher et al. 2003). This research (Bennett and Christie 2007; Bennett et al. 2006), which has led only to single case studies but not to a viable method for meteorological purposes so far, uses a fibre-optic based continuous-wave Doppler LIDAR. Infra red radiation at $\lambda = 1.55\ \mu m$ from a distributed feedback laser diode is split between a reference path and a signal path. Radiation following the signal path is amplified and transmitted towards a scattering target, whence some tiny proportion is returned to the instrument. This is then mixed in a coupler with delayed radiation that has followed the reference path. A narrow-band beat signal can only be expected when the length of the delay line matches the path length in the atmosphere to within the coherence length of the source. Scattering from other distances will give a beat signal widened by twice the linewidth of the source (Harris et al. 1998). In the given instrument, the line width of the source is $\Delta\nu = 1.4$ MHz; the coherence length for the beats signal is $\Delta r = c/(2\pi\Delta\nu) = 34$ m, and this should be the effective range resolution of the system. The range of the system is between 42 and 217 m. For a reliable operation the atmospheric visibility should be below 30 km.

3.5.5 *Other Optical Flow Measurement Techniques*

A further method to measure wind flow that is based on aerosol backscatter is laser Doppler anemometry (LDA, Durst et al. 1976), also sometimes called laser Doppler velocimetry (LDV), which relies on the interference of two coherent laser beams. Using a beam splitter, the instrument emits two coherent beams, which cross each other at some distance (probably much less than 100 m) from the device. At the position of the cross-over, an interference pattern is formed (the frequency of the resulting beats is also called Doppler frequency which explains the name of this measurement technique). Aerosol particles floating in the air passing through this cross-over region are alternately illuminated or in darkness. From the frequency of this intensity oscillation, the air speed perpendicular to the fringes of the interference pattern can be determined. Both velocity components can be derived with this technique by producing a second pattern turned by 90° with respect to the first one. Successful boundary layer application of this technique, which has been designed primarily for wind tunnel flow measurements, has not been reported. Most probably, the sensitivity at longer ranges is not sufficient for this purpose in the unseeded atmosphere.

Particle imaging velocimetry (PIV, Adrian 1991) is another method commonly used for flow measurement in wind tunnels. Like LDA, it requires careful seeding of the flow with particles or oil droplets and is therefore unsuitable for remote sensing in the atmosphere. The same comments also apply to the technique of Doppler Global Velocimetry (DGV, Meyers 1995).

3.6 Radiometers

Radiometers record passively the intensity and sometimes additionally the polarisation of the incident radiation. The use of radiometers in remote sensing aims to determine the spectral and physical characteristics of incoming radiation. By an appropriate choice of the detector material, the desired spectral sensitivity of the instrument can be achieved. Indium antimonid, for instance, is suited to the detection of near infrared radiation, while cadmium tellurid, on the other hand, is sensitive to longer infrared radiation (Strangeways 2003).

Radiation intensity is a function of the temperature of the emitting object as described by Planck's law. On the low frequency side ($h\nu/(kT) >> 1$) the Rayleigh-Jean approximation is valid, where

$$\mathrm{E}(\nu) = 2kT\ \nu^2/c^2, \tag{3.21}$$

here, ν is frequency, $k = 1.381 \times 10^{-23}\,\mathrm{J\,K^{-1}}$ is the Boltzmann constant, T is absolute temperature, and c is the speed of light. The so-called brightness temperature, T_β can be obtained from (3.21) by an inversion. For a black body (emissivity $\epsilon = 1$), brightness temperature equals the true temperature of the object. For lower emissivities, the brightness temperature is lower than the true temperature, where

$$T_\mathrm{b} = \varepsilon T. \tag{3.22}$$

Radiometers sited at the ground detect a brightness temperature depending on the zenith angle, Θ according to

$$T_\mathrm{b}(\Theta) = T_\mathrm{bg} \exp(-i(h_1, \Theta)) + \sec\Theta \int \alpha(h)\ T(h)\ \exp(-\tau(h, \Theta))\ \mathrm{d}h, \tag{3.23}$$

where the integral is taken over the whole depth of the atmosphere. Here, T_bg is the background temperature (the radiation temperature of space) and $\alpha(h)$ the absorption coefficient at a height, h. The optical depth, τ, of a layer up to the height, h, is determined from the integral over the absorption coefficient from the surface to this height as given by

$$\tau(h, \Theta) = \sec\Theta \int \alpha(h')\ \mathrm{d}h'. \tag{3.24}$$

A particular radiometer that scans the whole hemisphere above it, and whose data processing is based on (3.24), is described in detail in Martin et al. (2006).

Radiometers operating in the microwave spectral range between 5 and 80 GHz (microwave scanning radiometers, MWSR) are important for the detection of vertical profiles of temperature, humidity, and water droplets (Dabberdt et al. 2004), because they can be used day and night with profiling interrupted only by rain. Validations and enhancements are still ongoing (Rouffieux et al. 2006). In the microwave spectral range, the emissivity ϵ of the surface varies strongly with surface properties so that microwave radiometers on satellites detect mainly emissivity differences. This is in striking contrast to the infrared spectral range where the emissivity is nearly constant and the received signal is primarily dominated by temperature differences of the surface.

Major emission bands in the microwave region between 20 and 200 GHz are two hydrogen bands at 22 GHz and 183 GHz, a very broad oxygen band at 60 GHz, and an isolated oxygen line at 119 GHz (Cimini et al. 2006). Height attribution of the received signal is made from the pressure-related broadening of these lines and bands. In optically thick bands such as the oxygen band at 60 GHz, measurements may be made in different distances from the band centre. The transparency of the atmosphere increases with increasing distance from the band centre, so that the radiation received at a satellite stems from deeper and deeper layers. The exact location of the band centres is temperature-dependent.

A radiometer that is sensitive in the spectral range of visible light is called photometer. It is designed to measure the intensity of light sources. The detection is made with photo-diodes. The spectral sensitivity of these diodes is adapted to the sensitivity of the human eye by filters brought into the light beam. Photo-resistors made from cadmium sulphide can also be used (Strangeways 2003).

A novel development is HATPRO (Humidity And Temperature PROfiler HATPRO, Rose et al. (2005)) with high temporal resolution. Compared to other microwave radiometers, HATPRO is able to observe temperature profiles with high vertical resolution in the atmospheric boundary layer in addition to the standard products IWV (integrated water vapour), LWP (cloud liquid water path), and full troposphere temperature and humidity profiles. The enhanced vertical resolution in the boundary layer is achieved by conical scanning at six different elevation angles at a frequency where the atmosphere is quite opaque. HATPRO continuously measures thermal emission by atmospheric components (water vapour, oxygen, liquid water) simultaneously in 14 channels distributed over two frequency bands. Seven channels are located along the high-frequency wing of the water vapour absorption line at 22.235 GHz (22.24–31.4 GHz), and another 7 channels are located along the low-frequency wing of the 60 GHz oxygen absorption complex (51.26–58 GHz). HATPRO is operating in zenith pointing mode most of the time (except when performing the scans at six different elevation angles) giving a set of 14 brightness temperatures every 2 seconds. Statistical algorithms are used in order to retrieve atmospheric parameters from brightness temperatures. These algorithms are developed on the basis of synthetic observations generated from a representative long-term data set (Pospichal and Crewell 2007).

3.6.1 Infrared Interferometer

The Atmospheric Emitted Radiance Interferometer (AERI) is a high spectral resolution interferometer sounder, which records spectra in the infrared (3.3–18.2 μm) with a spectral resolution of better than 1 cm^{-1} at less than 10-min temporal resolution (Feltz et al. 2003). These radiance spectra are then converted into temperature and humidity profiles by an inversion of the radiation transfer equation (Knuteson et al. 2004a, b). This inversion requires a first guess, which may be taken from a radiosonde measurement or from climatology. A typical measurement cycle consists of a 3-min sky dwell period followed by 2-min dwell periods for each of the blackbodies. Although the interferometer acquires an uncalibrated spectrum every 2 s, averaging reduces the radiometric noise in the measurements. The instrument contains two calibration blackbodies, one at 60°C and one at ambient temperature, which allow for a self-calibration of the instrument (Feltz et al. 2003). The high spectral resolution of an interferometer allows for significantly improved accuracy and vertical resolution of retrieved temperature and humidity profiles. AERI can measure up to a height of about 3 km above ground. The vertical resolution is 100–200 m.

3.7 FTIR

The Fast Fourier Infrared (FTIR) absorption spectroscopy (Bacsik et al. 2004) in its active version is a bistatic method operating a heat source (broad-band infrared radiation source, called glo(w)bar, see Fig. 3.14) and a spectrometer based on a Michelson interferometer. Typical path lengths are several hundreds of metres. In a passive version the sun or another well-known heat source can be used. Passive FTIR emission spectroscopy analyses the emission of gases in front of a cooler background. The infrared detector of a FTIR needs cooling down to 77 K with liquid nitrogen or by the operation of a Joule-Thompson cooling system.

The interferometer in the FTIR spectrometer changes the incoming frequency-dependent radiation intensity $I(\nu)$ into its Fourier transform $\hat{I}(x)$. The following relations apply:

$$\hat{I}(x) = 1/(2\pi) \int_{-\infty}^{+\infty} I(\nu)\ \cos\,(2\pi\ \nu x)\, \mathrm{d}\nu \qquad (3.25)$$

and

$$I(\nu) = \int_{-\infty}^{+\infty} \hat{I}(x)\ \cos\,(2\pi\ \nu x)\, \mathrm{d}x. \qquad (3.26)$$

The great advantage of an interferometer compared to a spectrometer using a split or grate is the high light intensity, which reaches the detector. This lowers the signal-to-noise ratio by a factor of about 200. This advantage is also known as the Jacquinot advantage (Griffiths 1983). In order to convert the interferograms back into spectra (3.26) must be solved. An efficient and fast method to solve (3.26) was developed by Cooley and Tuckey (1965). The modern computer technology today permits a rapid online calculation of spectra from interferograms.

Further advantages of the FTIR spectroscopy are that all frequencies are observed simultaneously (at a grate spectrometer they are observed subsequently), which speeds up the observation up to a factor of 500 (multiplex or Fellget advantage (Fellgett 1951; Compton et al. 1990)). FTIR spectroscopy offers a high spectral resolution ($f/\Delta f \sim 100\ 000$, Hase and Fischer (2005). These advantages allow for a simultaneous detection of many trace gases along the measurement path.

Fig. 3.14 Glowbar as an infrared radiation source for open-path FTIR measurements

As a method working in the infrared FTIR is suited to detect trace gases which have emission or absorption lines in this spectral range. These are especially the greenhouse gases.

3.8 DOAS

The differential optical absorption spectroscopy (DOAS) method employs the fact that scattering and absorption depend differently on the light frequency. Mie and Rayleigh scattering only vary slowly with frequency while absorption varies rapidly. DOAS is an active method requiring a well-defined light source (lamp). In the bistatic version, lamp and receiving telescope are located several hundreds of meters up to a few kilometres from each other. In the monostatic version, lamp and telescope are in the same position and a passive retro-reflector is mounted in a respective distance (Fig. 3.15). DOAS was developed in the 1980s as a standard measurement

Fig. 3.15 DOAS instrumentation for open-path trace gas concentration measurements. *Left*: light source and receiving telescope, *right*: mirror. In operation the two parts are installed several hundreds of metres apart

method (Platt and Perner 1984). In contrast to the FTIR spectroscopy, the spectral resolution is lower ($f/\Delta f$ approximately some 100 up to some 1000, Hase and Fischer (2005)).

The DOAS light source should ideally emit white light. For this purpose xenon or halogen lamps are used in the focus of a mirror with 0.3 m diameter and 0.25 m focal length. In special cases, e.g. for the detection of OH radicals, dye lasers nay be employed as light source. The receiving telescope (focal length 0.25–0.85 m) focuses the light on a grate with 600–2160 lines per millimetre (Platt and Perner 1984). DOAS detects trace gases, which absorb in the UV or visible spectral range.

3.9 Scintillometry

Scintillometers use a bistatic method that analyses refractive index fluctuations influencing the propagation of optical (e.g. in the near infrared at 940 nm) or electromagnetic waves (e.g. microwaves at 94 GHz having a wave length of 3.2 mm) along the path from the emitter to the receiver (Fig. 3.16). This fluctuation phenomenon, which is called scintillation, is the same which makes stars twinkling at night. Two types of optical scintillometers are available: a small aperture or laser scintillometer (SAS, see, e.g. de Bruin 2002) and a large aperture scintillometer (LAS, see

Fig. 3.16 Scintillometer with emitting device in the background and receiving device in the foreground (Courtesy of Scintec AG, Germany)

e.g. Kleissl et al. (2008) for a LAS intercomparison study). The first operate with a thin laser beam, the latter with a wide light beam with a diameter of 10–20 cm, which is either produced by optical widening of a normal light beam with a diffusor or by a larger array of light-emitting diodes (LED). The more modern method with the LEDs makes it easier to produce a uniform light intensity over the whole beam width. The advantage of LAS compared to SAS is that LAS are not so much affected by a saturation of the signal along the path (for saturation effects see, e.g. Kohsiek et al. 2006). On the other hand, due to the large beam diameter, LAS in contrast to SAS cannot be used to determine the inner turbulence length scale, which is in the order of 10 mm and is a direct measure for the dissipation rate turbulent kinetic energy. While SAS and LAS are most sensitive to temperature fluctuations, microwave scintillometer (MWS) also register scintillation from water vapour fluctuations so that the combination of MWS and LAS can be used to derive latent heat fluxes in the surface layer (see, e.g. Beyrich et al. 2005).

The path length can be chosen from between several hundreds of metres to several kilometres, depending on the power of the emitter. The measurement yields the structural parameter ${C_n}^2$ (n = refractive index) related to the turbulence intensity of the air. This parameter may be used to estimate the magnitude (but not the direction) of vertical turbulent fluxes (see, e.g. Thiermann and Grassl (1992), or Beyrich et al. (2002)). A scintillometer must be placed in the atmospheric surface layer where Monin-Obukhov similarity theory applies. The scintillometer beam height and set-up distance have to be carefully chosen, such that the instrument operates in the weak scattering regime, where the structure parameter of the refractive index ${C_n}^2$ is small enough that it can be derived from the variance of signal intensity I from first-order scattering theory (Clifford et al. 1974).

Some scintillometers, called crosswind scintillometers, operate with two parallel beams in the horizontal (Lawrence et al. 1972). These instruments can estimate the cross-wind speed from the time-lagged cross-variance between the signals of the two beams as well. Together with the knowledge of the wind direction, thus the horizontal wind speed can be derived. Furthermore, if the surface roughness length is available, the surface layer similarity equations for the wind speed and ${C_n}^2$ can be solved iteratively to yield two prime turbulence variables: the friction velocity u_* and the surface flux of sensible heat. A sensitivity analysis suggests that, in optimum conditions, a crosswind scintillometer should be capable of providing u_* and the turbulent heat flux with uncertainties of 10–15% and 20–30%, respectively (Andreas 2000).

3.10 Tomography

Tomographic methods are able to deliver cross-sections through an air volume. It is often also called computed tomography (CT). The optical method is based on integrated optical measurements along a net of intersecting paths covering the area or volume of interest and a subsequent inverse computation of the absorption along the cross-sections. The optical emitters and receivers can be installed either along the perimeter of the area of interest (if the area to be probed should not be accessed) or the light source can also be erected inside the area with all the detectors at the perimeter. The latter mode of installation can be realised with less effort (Hashmonay et al. 1999; Wu et al. 1999). The original idea for this technique dates back to Byer and Shepp (1979). Tomographic methods are also applied in several other disciplines such as medicine, geophysics, or oceanography.

The acoustic travel time tomography is based on an analogue principle (Wilson and Thomson 1994; Raabe et al. 2001; Tetzlaff et al. 2002). Here, acoustic emitters and receivers are mounted on the perimeter of an area (Fig. 3.17). From the travel

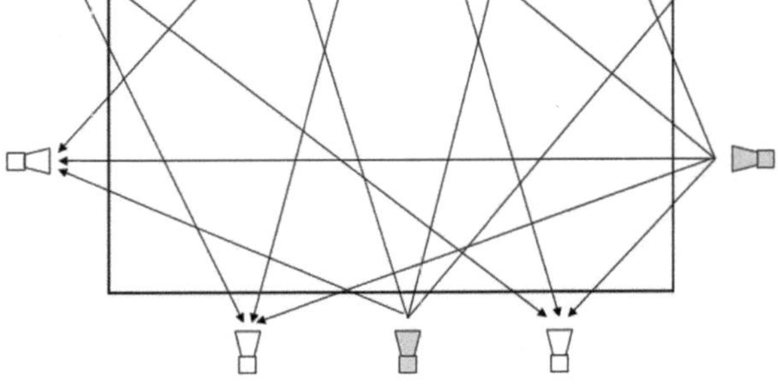

Fig. 3.17 Schematic design of the arrangement of sound sources (full symbols) and receivers (open symbols) for acoustic travel-time tomography of wind and temperature fields in a rectangular area

time of the sound pulses, the areal distribution of acoustic air temperature can be obtained. If the sound pulses are sent back and forth on all paths, the two wind components in the plane enclosed by the instruments can also be obtained. In this, tomography can be operated in analogue to a sonic anemometer. A recent review of acoustic tomography has been written by Ostashev et al. (2009).

Generally, stationary conditions are required until all measurements entering one inversion are taken. Optical tomography is obstructed by fog and strong precipitation, acoustic tomography by strong ambient noise and by heavy precipitation. Orographic features and buildings may prevent a sufficient coverage of the measurement area. Several evaluation techniques for tomographic measurements are available: the Simultaneous Iterative Reconstruction Technique (SIRT) algorithm (Humphreys and Clayton 1988; Trampert and Leveque 1990; Tetzlaff et al. 2002), Algebraic Reconstruction Technique (ART) (Todd and Leith 1990; Todd and Ramachandran 1994), and the Smooth Basis Function Minimization (SBFM) (Drescher et al. 1996). Recently, the time-dependent stochastic inversion (TDSI) technique was developed by Vecherin et al. (2006).

References

Abreu VJ, Barnes JE, Hays PB (1992) Observations of wind with an incoherent lidar detector. Appl Opt 31:4509–4514

Adrian RJ (1991) Particle-imaging techniques for experimental fluid mechanics. Ann Rev Fluid Mech 23:261–304

Al-Sakka H, Weill A, Le Gac C, Ney R, Chardenal L, Vinson J-P, Barthels L, Dupont E (2009) CURIE: a low power X-band, low atmospheric boundary layer Doppler radar. Meteorol Z 18:267–276

Andreas EL (2000) Obtaining surface momentum and sensible heat fluxes from crosswind scintillometers. J Atmos Oceanic Technol 17:3–16

Ansmann A, Riebesell M, Weitkamp C (1990) Measurement of atmospheric aerosol extinction profiles with a Raman lidar. Opt Lett 15:746–748

Bacsik Z, Mink J, Keresztury G (2004) FTIR spectroscopy of the atmosphere. I. Principles and methods. Appl Spectr Rev 39:295–363

Banakh VA, Smalikho IN, Köpp F, Werner C (1995) Representativeness of wind measurements with a cw Doppler lidar in the atmospheric boundary layer. Appl Opt 34:2055–2067

Bean BR, Dutton EJ (1968) Radio meteorology. Dover Publications, New York, 435 pp.

Behrendt A, Reichardt J (2000) Atmospheric temperature profiling in the presence of clouds with a pure rotational Raman lidar by use of an interference-filter-based polychromator. Appl Opt 39:1372–1378

Behrendt A, Nakamura T, Onishi M, Baumgart R, Tsuda T (2002) Combined Raman lidar for the measurement of atmospheric temperature, water vapor, particle extinction coefficient, and particle backscatter coefficient. Appl Opt 41:7657–7666

Bennett M, Christie S (2007) Doppler lidar measurements using a fibre optic system. Meteorol Z 16:469–477

Bennett M, Edner H, Grönlund R, Sjöholm M, Svanberg S, Ferrara R (2006) Joint application of Doppler lidar and differential absorption lidar to estimate the atomic mercury flux from a chlor-alkali plant. Atmos Environ 40:664–673

Beyrich F, de Bruin HAR, Meijninger WML, Schipper JW, Lohse H (2002) Results from one-year continuous operation of a large aperture scintillometer over a heterogeneous land surface. Bound-Lay Meteorol 105:85–97

Beyrich F, Kouznetsov RD, Leps J-P, Lüdi A, Meijninger WML, Weisensee U (2005) Structure parameters for temperature and humidity from simultaneous eddy-covariance and scintillometer measurements. Meteorol Z 14:641–649
Bingöl F (2005) Adapting a Doppler laser anemometer to wind energy. Masters thesis online, http://www.fm.mek.dtu.dk/upload/institutter/mek/fm/eksamensprojekter/ferhat%20bingöl%20mek-fm-ep%202005-02.pdf, accessed on 12 August 2010
Bonner CS, Ashley MCB, Lawrence JS, Storey JWV, Luong-Van DM, Bradley SG (2008) Snodar: a new instrument to measure the height of the boundary layer on the Antarctic plateau. Proceedings of SPIE 7014, 70146I, 7 pp., http://spie.org
Bragg WH, Bragg WL (1913) The reflection of X-rays by crystals. Proc Roy Soc Lond A 88:428–438
Byer RL, Shepp LA (1979) Two-dimensional remote air pollution monitoring via tomography. Opt Lett 4:75–77
Caccia J-L, Guénard V, Benech B, Campistron B, Drobinski P (2004) Vertical velocity and turbulence aspects during Mistral events as observed by UHF wind profilers. Annales Geophysicae 22:3927–3936
Canin ML, Garnier A, Hauchecorne A, Porteneuve J (1989) A Doppler lidar for measuring winds in the middle atmosphere. Geophys Res Lett 16:1273
Chan PW, Shun CM, Wu KC (2006) Operational LIDAR-based system for automatic windshear alerting at the Hong Kong International Airport. 12th Conference Aviation, Range, and Aerospace Meteorology, Paper 6.11, American Meteorological Society, Boston, MA, USA
Cimini D, Hewison TJ, Martin L, Güldner J, Gaffard C, Marzano FS (2006) Temperature and humidity profile retrievals from ground-based microwave radiometers during TUC. Meteorol Z 15:45–56
Clifford S, Ochs G, Lawrence R (1974) Saturation of optical scintillation by strong turbulence. J Opt Soc Amer 64:148–154
Colls J (2002) Air Pollution. Clay's Library of Health and the Environment, 2nd edn. Spon Press, London, 560 pp.
Compton DAC, Drab J, Barr HS (1990) Accurate infrared transmittance measurements on optical filters using an FT-IR spectrometer. Appl Opt 29:2908–2912
Cooley CW, Tuckey JW (1965) An algorithm for the machine calculation of complex Fourier series. Math Comput 19:297–301
Cooney J (1970) Remote measurements of atmospheric water-vapour profiles using the Raman component of laser backscatter. J Appl Meteorol 9:182–184
Cooney J (1972) Measurement of atmospheric temperature profiles by Raman backscatter. J Appl Meteorol 11:108–112
Dabberdt WF, Frederick GL, Hardesty RM, Lee W-C, Underwood K (2004) Advances in meteorological instrumentation for air quality and emergency response. Meteorol Atmos Phys 87:57–88
Davies F, Collier CG, Bozier KE, Pearson GN (2003) On the accuracy of retrieved wind information from Doppler lidar observations. Quart J Roy Meteor Soc 129:321–334
de Bruin HAR (2002) Introduction: renaissance of scintillometry. Bound-Lay Meteorol 105:1–4
Drescher AC, Gadgil AJ, Price PN, Nazaroff WW (1996) Novel approach for tomographic reconstruction of gas concentration distributions in air: use of smooth basis functions and simulated annealing. Atmos Environ 30:929–940
Durst F, Melling A, Whitelaw JH (1976) Principles and Practice of Laser Doppler Anemometry, Academic Press, London
Ecklund WL, Carter DA, Balsley BB (1988) A UHF wind profiler for the boundary layer: brief description and initial results. J Atmos Oceanic Technol 5:432–441
Emeis S (2010) Measurement Methods in Atmospheric Sciences – In situ and Remote. Borntraeger, Stuttgart, 257 pp.
Emeis S, Harris M, Banta RM (2007) Boundary-layer anemometry by optical remote sensing for wind energy applications. Meteorol Z 16:337–347

Engelbart DAM, Bange J (2002) Determination of boundary-layer parameters using wind profiler/RASS and sodar/RASS in the frame of the LITFASS project. Theor Appl Climatol 73:53–65

Fellgett PB (1951) Theory of infra-red sensitivites and its application to investigations of stellar radiation in the near infra-red. PhD Thesis, Cambridge University Press, Cambridge (UK)

Feltz WF, Smith WL, Howell HB, Knuteson RO, Woolf H, Revercomb HE (2003) Near-continuous profiling of temperature, moisture, and atmospheric stability using the Atmospheric Emitted Radiance Interferometer (AERI). J Appl Meteorol 42:584–597

Fercher AF, Drexler W, Hitzenberger CK, Lasser T (2003) Optical coherence tomography – principles and applications. Rep Progr Phys 66:239–303

Fiocco G, Smullin LD (1963) Detection of scattering layers in the upper atmosphere /60–140 km/ by optical RADAR. Nature 199:1275–1276

Gilman GW, Coxhead HB, Willis FH (1946) Reflection of sound signals in the troposphere. J Acoust Soc Amer 18:274–283

Griffiths PR (1983) Fourier transform infrared spectroscopy. Science 222:297–302

Grund CJ, Banta RM, George JL, Howell JN, Post MJ, Richter RA, Weickmann AM (2001) High-resolution Doppler lidar for boundary layer and cloud research. J Atmos Oceanic Technol 18:376–393

Hannon SM (2004) Pulsed Doppler lidar for terminal area monitoring of wind and wake hazards. Proceedings of the 11th Conference Aviation, Range, and Aerospace Meteorology Paper P4.21, American Meteor Soc

Hardesty RM, Darby LS (2005) Ground-based and airborne lidar. In: Anderson MG (ed.) Encyclopedia of Hydrologic Sciences. Wiley, New York, 697–712

Harris M, Pearson GN, Vaughan JM, Letalick D, Karlsson C (1998) The role of laser coherence length in continuous-wave coherent laser radar. J. Modern Optics 45:1567–1581

Harris M, Constant G, Ward C (2001) Continuouswave bistatic laser Doppler wind sensor. Appl Opt 40:1501–1506

Harris M, Hand M, Wright A (2006) Lidar for Turbine Control. NREL, Golden, Colorado, Techn Rep NREL/TP-500-39154. vi + 47 pp. (available from http://www.nrel.gov/wind/pdfs/39154.pdf, accessed on 12 August 2010)

Hase F, Fischer H (2005) Satellitengestützte Fernerkundung atmosphärischer Spurenstoffe. promet 31:38–43

Hashmonay RA, Yost MG, Wu C-F (1999) Computed tomography of air pollutants using radial scanning path-integrated optical remote sensing. Atmos Environm 33:267–274

Humphreys E, Clayton RW (1988) Adaption of back projection tomography to seismic travel time problems. J Geophys Res 93:1073–1085

IEC 60825-1 Safety of laser products – Part 1: Equipment classification, requirements and user's guide. Geneva

Ince T, Frasier SJ, Muschinski A, Pazmany AL (2003) An S-band frequency-modulated continuous-wave boundary layer profiler: Description and initial results. Radio Sci 38:1072

Keckhut P, Chanin ML, Hauchecorne A (1990) Stratosphere temperature measurement using Raman lidar. Appl Opt 29:5182–5186

Kindler D, Oldroyd A, MacAskill A, Finch D (2007) An eight month test campaign of the Qinetiq ZephIR system: Preliminary results. Meteorol Z, 16:479–489

Kleissl J, Gomez J, Hong S-H, Hendrickx JMH, Rahn T, Defoor WL (2008) Large aperture scintillometer intercomparison study. Bound-Lay Meteorol 128:133–150

Knight CA, Miller LJ (1998) Early radar echoes from small, warm cumulus: Bragg and hydrometeor scattering. J Atmos Sci 55:2974–2992

Knuteson RO, Best FA, Ciganovich NC, Dedecker RG, Dirkx TP, Ellington S, Feltz WF, Garcia RK, Herbsleb RA, Howell HB, Revercomb HE, Smith WL, Short JF (2004a) Atmospheric Emitted Radiance Interferometer (AERI): Part I: Instrument Design. J Atmos Oceanic Technol 21:1763–1776

Knuteson RO, Best FA, Ciganovich NC, Dedecker RG, Dirkx TP, Ellington S, Feltz WF, Garcia RK, Herbsleb RA, Howell HB, Revercomb HE, Smith WL, Short JF (2004b) Atmospheric

Emitted Radiance Interferometer (AERI): Part II: Instrument Performance. J Atmos Oceanic Technol 21:1777–1789
Kohsiek W, Meijninger WML, de Bruin HAR, Beyrich F (2006) Saturation of the large aperture scintillometer. Bound-Lay Meteorol 121:111–126
Kouznetsov R (2009) The multi-frequency sodar with high temporal resolution. Meteorol Z 18:169–173
Lawrence RS, Ochs GR, Clifford SF (1972) Use of scintillation to measure average wind across a light beam. Appl Opt 11:239–243
Marshall JS, Palmer SM (1948) The distribution of raindrops with size. J Meteorol 5:165–166
Marshall JM, Peterson AM, Barnes AA (1972) Combined radar-acoustic sounding system. Appl Opt 11:108–112
Martin L, Schneebeli M, Mätzler C (2006) ASMUWARA, a ground-based radiometer system for tropospheric monitoring. Meteorol Z 15:11–17
Menut L, Flamant C, Pelon J, Flamant PH (1999) Urban boundary-layer height determination from lidar measurements over the Paris area. Appl Optics 38:945–954
Meyers JF (1995) Development of Doppler global velocimetry as a flow diagnostics tool. Meas Sci Technol 6:769–783
Münkel C, Emeis S, Müller WJ, Schäfer K (2003) Observation of aerosol in the mixing layer by a ground-based lidar ceilometer. In: Schaefer K, Lado-Bordowsky O, Comeron A, Picard RH (eds), Remote Sensing of Clouds and the Atmosphere VII, Proceedings of SPIE, 4882:344–352, http://spie.org
Münkel C, Eresmaa N, Räsänen J, Karppinen A (2007) Retrieval of mixing height and dust concentration with lidar ceilometer. Bound-Lay Meteorol 124:117–128
Ostashev VE, Vecherin SN, Wilson KD, Ziemann A, Goedecke GH (2009) Recent progress in acoustic travel-time tomography of the atmospheric surface layer. Meteorol Z 18:125–133
Peters G (1991) SODAR – ein akustisches Fernmeßverfahren für die untere Atmosphäre. promet 21:55–62
Peters G, Fischer B, Andersson T (2002) Rain observations with a vertically-looking Micro Rain Radar (MRR). Boreal Env Res 7:353–362
Peters G, Fischer B, Münster H, Clemens M, Wagner A (2005) Profiles of rain drop size distributions as retrieved by Micro Rain Radars. J Appl Met 44:1930–1949
Platt U, Perner D (1984) Ein Instrument zur spektroskopischen Spurenstoffmessung in der Atmosphäre. Fresenius Z Anal Chem 317:309–313
Pospichal B, Crewell S (2007) Boundary layer observations in West Africa using a novel microwave radiometer. Meteorol Z 16:513–523
Post MJ, Cupp RE (1990) Optimizing a pulsed Doppler lidar. Appl Opt 29:4145–4158
Raabe A, Arnold K, Ziemann A (2001) Near surface spatially averaged air temperature and wind speed determined by acoustic travel time tomography. Meteorol Z 10:61–70
Reitebuch O (1999) SODAR-Signalverarbeitung von Einzelpulsen zur Bestimmung hochaufgelöster Windprofile. Schriftenreihe Fraunhofer-Institute for Atmospheric Environmental Research, Shaker Verlag GmbH Aachen, Vol. 62, 178 pp.
Rose T, Crewell S, Löhnert U, Simmer C (2005) A network suitable microwave radiometer for operational monitoring of the cloudy atmosphere. Atmos Res 75:183–200
Rouffieux D, Nash J, Jeannet P, Agnew JL (2006) The COST 720 temperature, humidity and cloud profiling campaign TUC. Meteorol Z 15:5–10
Smith DA, Harris M, Coffey AS, Mikkelsen T, Jørgensen HE, Mann J, Danielian R (2006) Wind lidar evaluation at the Danish wind test site in Høvsøre, Wind Energy 9:87–93
Strangeways I (2003) Measuring the natural environment, second edn. Cambridge University Press, Cambridge 534 pp.
Tatarskii VI (1971) The effect of the turbulent atmosphere on wave propagation. Kefer Press, Jerusalem, 472 pp.
Tetzlaff G, Arnold K, Raabe A, Ziemann A (2002) Observations of area averaged near-surface wind- and temperature-fields in real terrain using acoustic travel time tomography. Meteorol Z 11:273–283

Thiermann V, Grassl H (1992) The measurement of turbulent surface-layer fluxes by use of bichromatic scintillation. Bound-Lay Meteorol 58:367–389

Todd L, Leith D (1990) Remote sensing and computed tomography in industrial hygiene. Amer Ind Hyg Ass J 51:224–233

Todd L, Ramachandran N (1994) Evaluation of algorithms for tomographic reconstruction of chemical concentrations in indoor air. Amer Ind Hyg Ass J 55:403–417

Trampert J, Leveque J-J (1990) Simultaneous iterative reconstruction technique: Physical interpretation based on the generalized least squares solution. J Geophys Res 95:12553–12559

Turner DD, Ferrare RA, Brasseur LAH, Feltz WF, Tooman TP (2002) Automated retrievals of water vapor and aerosol profiles from an operational Raman lidar. J. Atmos Oceanic Technol 19:37–50

Vecherin SN, Ostashev VE, Goedecke GH, Wilson DK, Voronovich AG (2006) Time-dependent stochastic inversion in acoustic travel-time tomography of the atmosphere. J Acoust Soc Am 119:2579–2588

VDI 3786 Part 14 (2001) Environmental meteorology – ground-based remote sensing of the wind vector – Doppler wind LIDAR. Beuth, Berlin

VDI 3786 Part 17 (2007) Environmental meteorology – ground-based remote sensing of the wind vector – wind profiler radar. Beuth, Berlin

Vogt S, Wieser A (2006) Doppler wind lidar and radar wind profiler – comparison of instruments and discussion of various measurement methods. Ext Abstr Intern Symp for the Advancement of Boundary Layer Remote Sensing (ISARS 13), 18–20 July 2006, Garmisch-Partenkirchen, Germany. Wiss Ber FZKA 7222:53–55

Wächter M, Rettenmeier A, Kühn M, Peinke J (2009) Characterization of short time fluctuations in atmospheric wind speeds by LIDAR measurements. Meteorol Z, 18:277–280

Weitkamp C (ed.) (2005) Lidar. Range-Resolved Optical Remote Sensing of the Atmosphere. Springer Science + Business Media Inc., New York, 455 pp.

Wexler R, Swingle DM (1946) Radar storm detection. Bull Amer Meteor Soc 28:159–167

Wilson DK, Thomson DW (1994) Acoustic tomography monitoring of the atmospheric surface layer. J Atmos Ocean Tech 11 751–769

Wu C-F, Yost MG, Hashmonay RA (1999) Experimental evaluation of a radial beam geometry for mapping air pollutants using optical remote sensing and computed tomography. Atmos Environ 33:4709–4716

Wulfmeyer V (1998) Ground-based differential absorption lidar for water-vapor and temperature profiling: Requirements, development, and specifications of a high-performance laser transmitter. Appl Opt 37:3804–3824

Chapter 4
Applications

4.1 Introduction

This chapter gives an overview of some characteristic applications of surface-based remote-sensing instruments, which have been described in the previous chapter, in the field of boundary layer meteorology. Today, results from boundary layer remote sensing are used, e.g. for basic research on vertical and horizontal exchange processes in the boundary layer, for air quality assessment, and for energy meteorology. The spectrum of application ranges from qualitative measurements for the detection of the overall vertical layering of the ABL (Section 4.2) over quantitative measurements of the vertical distribution of a variety of atmospheric parameters (Section 4.3) and fluxes (Section 4.4) to the investigation of certain three-dimensional flow systems in the ABL (Section 4.5).

The selected examples give an impression of the present state and abilities of experimental boundary layer research. Being examples, they do not represent a complete overview of all possible applications. This is not only due to the limited space in the present book but also due to the still rapidly ongoing developments in this field. The outlook in the concluding chapter, which follows after this one, will try to shed light on some of the most recent achievements.

Most of the presented applications refer to vertical profiling and sounding. This is because the state of the atmospheric boundary layer is mainly determined by its vertical structure and the vertical gradients of the atmospheric variables in this layer. Nevertheless, horizontal remote sensing such as tomography, scintillometry, DOAS, and FTIR will be mentioned where appropriate.

4.2 Vertical Layering of the ABL

The vertical layering of the ABL has an important influence on the atmosphere–geosphere interaction. Here, geosphere comprises the sphere of human activities, the biosphere, hydrosphere, cryosphere, and the solid Earth. The state of the ABL is crucial for exchange processes between the geosphere and the atmosphere as the layering of the ABL enables, prevents, or modifies vertical and horizontal transports

S. Emeis, *Surface-Based Remote Sensing of the Atmospheric Boundary Layer*, Atmospheric and Oceanographic Sciences Library 40, DOI 10.1007/978-90-481-9340-0_4,

of energy, momentum, moisture, or trace substances. Therefore, the detection of the vertical layering is one of the principal tasks of experimental boundary layer research.

We must distinguish between the mixing layer height, MLH (see Section 4.2.1) and the boundary layer height, z_i (see Section 4.2.2). The boundary layer height is the height up to which the influence of the presence of the lower surface is detectable. The mixing layer height is the height up to which atmospheric properties or substances originating from the surface are dispersed by turbulent vertical mixing processes. The mixing layer – if it is present at all – is a part of the atmospheric boundary layer. Thus, the mixing layer height is usually shallower than the boundary layer, but it fills the whole ABL in deep convective boundary layers.

4.2.1 Mixing Layer Height

The mixing layer height is the height up to which atmospheric properties or substances originating from the Earth's surface or formed within this layer are dispersed almost uniformly over the entire depth of this layer by turbulent vertical mixing processes. Therefore, the existence and the height of a mixing layer can be analyzed either from a detection of the presence of the mixing process, i.e. turbulence, or from the verification that a given conservative atmospheric variable is distributed evenly over a certain height range. The level of turbulence can, for instance, be derived from fluctuations of the wind components or from temperature fluctuations. Suitable conservative atmospheric variables for the identification of the mixing layer and its heights are, e.g. potential temperature, specific humidity or aerosol particle concentrations.

A rather complete overview of methods to determine the MLH from in situ measurements and surface-based remote sensing has been given by Seibert et al. (2000). Since then considerable development has taken place, especially concerning the usage of surface-based remote-sensing methods (see the review paper by Emeis et al. (2008)). This subchapter will mainly follow this latter review.

Optical methods for MLH detection may be used to illustrate this recent progress. Seibert et al. (2000) still classified LIDAR methods as expensive, not eye-save, with a high lowest range gate, limited range resolution, and sometimes subject to ambiguous interpretation. This has changed drastically in the past ten years when better LIDARs have been built and ceilometers have been discovered to be a nearly ideal boundary layer sounding instrument. Progress has been made in the field of acoustic sounding as well. Similarly, algorithms for the determination of MLH from vertical profiles of the acoustic backscatter intensity as described in Beyrich (1997) and Seibert et al. (2000) have been enhanced by using further variables available from SODAR measurements such as the wind speed and the variance of the vertical velocity component (Asimakopoulos et al. 2004, Emeis and Türk 2004). Such enhancements had been named as possible methods in Beyrich (1995) and Seibert et al. (2000), but obviously no example was available at that time.

Table 4.1 Overview on methods using ground-based remote sensing for the derivation of the mixing layer height mentioned in this chapter (see rightmost column for section title)

Method	Short description	Section
Acoustic ARE	Analysis of acoustic backscatter intensity	*Acoustic Received Echo Method*
Acoustic HWS	Analysis of wind speed profiles	*Horizontal Wind Speed Method*
Acoustic VWV	Analysis of vertical wind variance profiles	*Vertical Wind Variance Method*
Acoustic EARE	Analysis of backscatter and vertical wind variance profiles	*Enhanced Acoustic Received Echo Method*
Optical threshold	Detection of a given backscatter intensity threshold	*Threshold Method*
Optical gradient	Analysis of backscatter intensity profiles	*Gradient or Derivative Methods*
Optical idealized backscatter	Analysis of backscatter intensity profiles	*Idealized Backscatter Method*
Optical wavelet	Analysis of backscatter intensity profiles	*Wavelet Method*
Optical variance	Analysis of backscatter intensity profiles	*Variance Method*
Acoustic/electromagnetic	RASS	RASS
Acoustic/electromagnetic	SODAR-RASS and windprofiler-RASS	*Combined Deployment of Two Different RASS*
Acoustic/in situ	SODAR-RASS plus surface heat flux data	*Further Algorithms Using a RASS*
Acoustic/electromagnetic	SODAR plus windprofiler	*Combined Deployment of SODAR and Wind Profiler*
Acoustic/optical	SODAR plus ceilometer	*Combined Deployment of SODAR and Ceilometer*

A variety of different algorithms have been developed by which the MLH is derived from ground-based remote-sensing data (see Table 4.1 for a short overview). We will mainly concentrate on acoustic and optical remote sensing because electromagnetic remote sensing has too high lowest range gates for a good coverage of shallow MLH. The disadvantage of a too high lowest range gate can partly be circumvented by slantwise profiling or conical scanning if the assumption of horizontal homogeneity can be made.

4.2.1.1 Acoustic Methods

Acoustic methods either analyze the acoustic backscatter intensity, or, if Doppler shifts in the backscattered pulses can be analyzed, features of vertical profiles of the wind components and its variances as well. The acoustic backscatter intensity is proportional to small-scale fluctuations in atmospheric temperature (usually generated by turbulence) or by stronger vertical temperature gradients. The latter feature

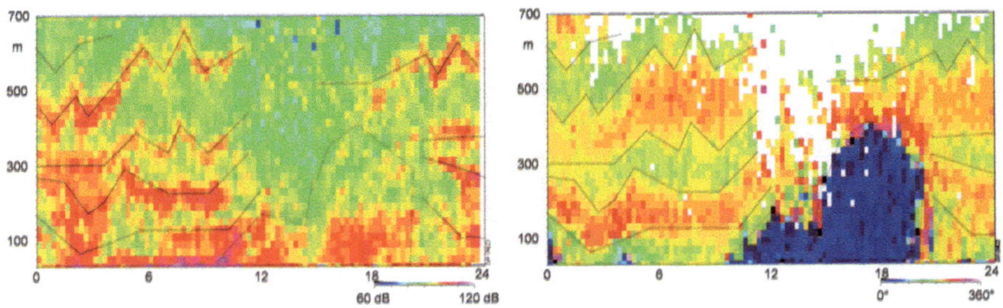

Fig. 4.1 Multiple layering of the atmosphere in a wintry Alpine valley (Inn valley near Schwaz, Austria) detected from SODAR measurements on January 29, 2006 (time is given in GMT+1). *Left*: acoustic backscatter intensity in dB (*green*: low, *red*: high), *right*: horizontal wind direction in degree (*blue*: around 20°, *greenish*: 190°, *reddish*: 230°). Black lines indicate layer boundaries determined with the EARE algorithm. Only the lowest layer, present between 10 pm and midnight, may be addressed as a mixing layer

may be an indication for the presence of temperature inversions, which can often be found at the top of the mixing layer.

Beyrich (1997) listed possible analyses that can mainly be made from acoustic backscatter intensities measured by a SODAR. Later, Asimakopoulos et al. (2004) summarized three different methods to derive MLH from SODAR data: (1) the horizontal wind speed method (HWS), (2) the acoustic received echo method (ARE), and (3) the vertical wind variance method (VWV). We will mainly follow this classification here and finally add a fourth method, the enhanced ARE method (EARE), in section "Enhanced Acoustic Received Echo Method".

Figure 4.1, showing an acoustic sounding taken in an Alpine valley, gives an impression what wealth of detailed vertical information can be derived from acoustic boundary layer sounding. The left-hand frame displays the acoustic backscatter intensity and the right-hand frame the wind direction as time-height sections over one day (from midnight to midnight) and over a height range of 700 m. The depicted wintry situation from a day in January exhibits a multiple layering of the air in that valley due to the very stable thermal stratification of the valley air over a snow-covered valley floor. The multiple layering originated from an interlacing of down-valley (wind direction around 190°) and down-slope (wind direction around 230°) flows. The layers are separated by temperature inversions, and each higher layer is potentially warmer than the next lower layer. They persisted nearly the whole day because no vertical mixing took place in the stably stratified valley atmosphere.

Acoustic Received Echo Method

The ARE method is the oldest and most basic method of determining MLH from acoustic remote sensing. Most of the methods listed in Beyrich (1997) belong to this method. The method does not require an analysis of the Doppler shift of the backscattered signals, but is based on the analysis of facsimile plots, i.e. time-height cross sections of the backscatter intensity. The method makes use of the assumption that turbulence is larger in the mixing layer than in the atmosphere above, and that this turbulence is depicted in enhanced intensity of the acoustic backscatter. MLH is analyzed either from the maximum negative slope or from the changing curvature of the vertical profile of the acoustic backscatter intensity, or it is analyzed from the height where the backscatter intensity decreases below a certain pre-specified threshold value.

Horizontal Wind Speed Method

The HWS method requires the analysis of the Doppler shift of the backscattered acoustic signals. The algorithm is based on the analysis of the shape of hourly-averaged vertical wind speed profiles using the assumption that wind speed and wind direction are almost constant within the mixing layer but approach gradually towards the geostrophic values above the mixing layer. Beyrich (1997) listed this method in his Table 2, but did not discuss it further. The applicability of the method is probably limited to well-developed convective boundary layers (CBL) due to the underlying assumptions. Such CBL are often higher than the maximum range of a SODAR. Even if the CBL height is within the range of the SODAR, the algorithm for the analysis of the Doppler shift often fails above the inversion topping the CBL due to too low signal-to-noise ratios.

Vertical Wind Variance Method

The VWV method is also working only for CBLs. It is based on the vertical profile of the variance of the vertical velocity component σ_w. In a CBL σ_w reaches a maximum in a height $a{\cdot}z_i$. Typical values for a are between 0.35 and 0.4. Thus, in principle, this is an extrapolation method. It has been tried for SODAR measurements because it permits a detection of MLH up to heights that are 2.5 times above the limited maximum range (usually between 500 and 1000 m) of the SODAR. Beyrich (1997) classified this method as not reliable. A related method, which is based on power spectra of the vertical velocity component, is integrated in the commercial evaluation software of certain SODARs (Contini et al. 2009).

Enhanced Acoustic Received Echo Method

The EARE algorithm has been proposed by Emeis and Türk (2004) and Emeis et al. (2007b). The method is an enhancement of the ARE method in two ways. Firstly, it includes further variables into the MLH algorithm that are available from Doppler-SODARs. The benefits of the additional usage of the variance of the vertical velocity component have been demonstrated by Emeis and Türk (2004). Secondly, it determines not only MLH from SODAR measurements but also the heights of additional lifted inversions. Especially in orographically complex terrain, the vertical structure of the ABL can be very complicated. Emeis et al. (2007a) have shown that several persistent inversions one above the other which form in deep Alpine valleys can be detected from SODAR measurements (Fig. 4.1).

EARE determines three different types of heights based on acoustic backscatter intensity and the variance of the vertical velocity component. Because the horizontal wind information above the inversion is not regularly available from SODAR measurements, horizontal wind data have not been included into this scheme. In the following, a letter "H" and an attached number will denote certain derived heights that are related to inversions and the MLH, while the variable z is used to denote the normal vertical coordinate. The EARE algorithm detects

- the height ($H1$) of a turbulent layer characterized by high acoustic backscatter intensities $R(z)$ due to thermal fluctuations (therefore having a high variance of the vertical velocity component σ_w),
- several lifted inversions ($H2\_n$) characterized by secondary maxima of acoustic backscatter due to a sharp increase of temperature with height and simultaneously low σ_w (like those depicted in the left-hand frame of Fig. 4.1), and
- the height of a surface-based stable layer ($H3$) characterized by high backscatter intensities due to a large mean vertical temperature gradient starting directly at the ground and having a low variance of the vertical velocity component.

The height $H1$ corresponds to a sharp decrease $\partial R/\partial z < DR_1$ of the acoustic backscatter intensity $R(z)$ below a threshold value R_c with height z usually indicating the top of a turbulent layer:

$$\begin{aligned} &H1 = z, \\ &\text{if } R(z) < R_c \text{ and } R(z+1) < R(z) + zDR_1 \text{ and } R(z+2) < R(z) + 2zDR_1, \end{aligned} \tag{4.1}$$

$R_c = 88$ dB and $DR_1 = -0.16$ dB/m have proven to be meaningful values in the abovementioned studies. R_c is somewhat arbitrary because the received acoustic backscatter intensities from a SODAR cannot be absolutely calibrated. An absolute calibration would require the knowledge of temperature and humidity distributions along the sound paths for a precise calculation of the sound attenuation in the air. DR_1 is, at least for smaller vertical distances, independent from the absolute value of R_c. An application-dependent fine-tuning of R_c and DR_1 may be necessary.

Elevated inversions are diagnosed from secondary maxima of the backscatter intensity that are not related to high turbulence intensities. For elevated inversions increase in backscatter intensity below a certain height $z = H2$ and a decrease above is stipulated while the turbulence intensity is low:

$$H2\_n = z, \text{if } \left.\frac{\partial R}{\partial z}\right|_{z+1} < -DR_2 \text{ and } \left.\frac{\partial R}{\partial z}\right|_{z-1} < DR_2 \text{ and } \sigma_w(z) < 0.7 \text{ ms}^{-1}, \tag{4.2}$$

for $n = 1, \ldots, N$. In Emeis et al. (2007b) N was chosen to be five. A threshold value $DR_2 = 0.08$ dB/m has proven suitable. But again, an application-dependent tuning may be advisable.

The determination of the height of the stable surface layer $H3$ is started if the backscatter intensity in the lowest range gates is above 105 dB while σ_w is smaller than 0.3 m s^{-1}. The top of the stable layer *H*3 is at the height where either the backscatter intensity sinks below 105 dB or σ_w increases above 0.3 m s^{-1}.

$$H3 = z, \text{if } (R(z) > 105\,dB \text{ and } R(z+1) < 105\,dB \text{ and } \sigma_w(z) < 0.3\ \text{ms}^{-1}) \text{ or if } (\sigma_w(z) < 0.3\ \text{ms}^{-1} \text{ and } \sigma_w(z+1) > 0.3\ \text{ms}^{-1} \text{ and } R(z) > 105\,dB), \quad (4.3)$$

The σ_w values used in eqs. (4.2) and (4.3) have been determined by optimizing the automatic application of the detection algorithm. In doing so it turned out that no lifted inversions occurred with a variance σ_w higher than 0.7 m s^{-1} and that the variance σ_w in nocturnal stable surface layers was below 0.3 m s^{-1}. The first σ_w threshold made it possible to distinguish between inversions and elevated layers of enhanced turbulence. The latter σ_w threshold made it possible to differentiate between nocturnal stable surface layers and daytime super-adiabatic surface layers although both types of surface layers yield more or less the same level of backscatter intensity. Finally, MLH from acoustic remote sensing is determined as the minimum of *H*1, *H*2_1, and *H*3:

$$\text{MLH}_{\text{ac}} = \min(H1, H2\_1, H3), \quad (4.4)$$

Two examples for the derivation of MLH from SODAR data are presented schematically in Fig. 4.2. On the left-hand side, a convective daytime situation is depicted. The high acoustic backscatter intensity near the ground is caused by the high turbulence due to the heating of the ground by the incoming shortwave solar radiation. This is supported by the simultaneously high standard deviation of the vertical wind component. Above this super-adiabatic layer a secondary maximum in acoustic backscatter intensity occurs, which is related to a local minimum in the standard deviation of the vertical wind component. This is a clear indication for the presence of a stable layer with low turbulence, i.e. an inversion. Therefore, this height is selected as MLH in this case.

On the right-hand side a nocturnal situation is presented. The high acoustic backscatter is here caused by a strong vertical increase of temperature in the stable nocturnal boundary layer. Once again, the low standard deviation of the vertical wind component supports this analysis. Therefore, MLH is analyzed to be at the top of this stable layer. The right-hand plot shows a secondary maximum aloft as well. But in this case it is coupled with high turbulence (high standard deviation of the vertical wind component), so it not selected as a stable layer which limits the vertical dispersion of pollutants.

4.2.1.2 Optical Detection of MLH

Usually the aerosol content of the mixing layer is higher than in the atmospheric layer above, because the emission sources for aerosols are in most cases on the ground. Aerosol formation from precursors mainly takes place near the surface as well. Making the assumption that the vertical aerosol distribution adapts rapidly to the changing thermal structure of the boundary layer, MLH can be determined

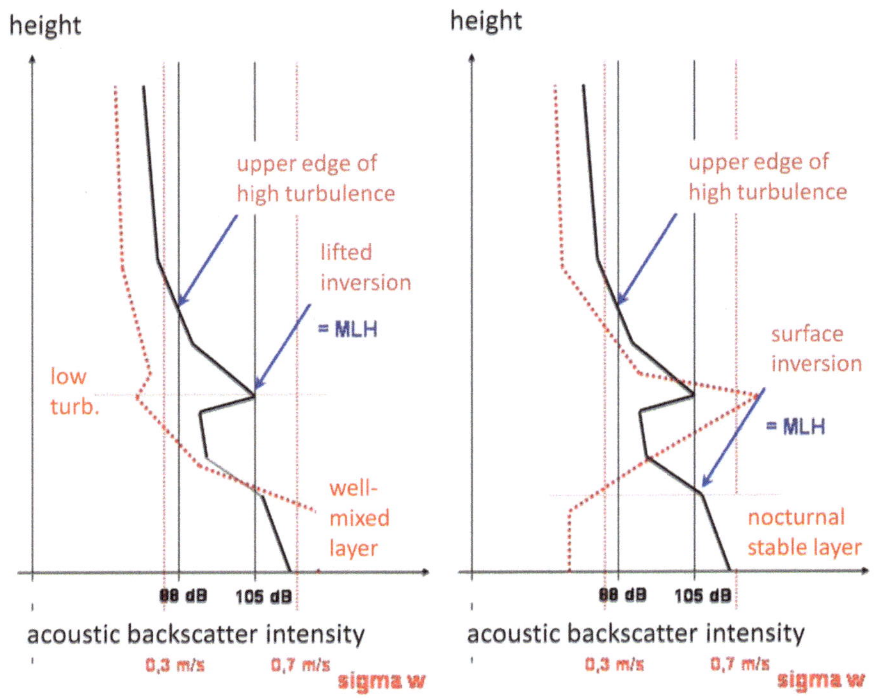

Fig. 4.2 Schematic of the determination of MLH from SODAR measurements. *Full black curve*: acoustic backscatter intensity in dB, *dotted red curve*: vertical profile of σ_w. *Left*: example for a daytime convective boundary layer, *right*: example for a stable nocturnal boundary layer. Vertical *black* and *red lines* indicate threshold values used in eqs. (4.1) to (4.3)

from the analysis of the vertical aerosol distribution. This also includes the assumption that the vertical aerosol distribution is not dominated by horizontally advected aerosol plumes or layers. The heights of the near surface aerosol layers (*H4_n*) can be analyzed from the optical vertical backscatter profile obtained from optical remote sensing. Several methods have been developed, the most prominent of these being: (1) the threshold method, (2) the gradient or derivative method, (3) the idealized gradient method, (4) the wavelet method, and (5) the variance method.

The application of optical remote sensing for MLH determination has focussed on the use of ceilometers in recent years. Ceilometers can be regarded as a small LIDAR. They are simpler and they have a much lower lowest range gate than LIDARs. For the detection of MLH below 150–200 m, a ceilometer with one optical axis for the emitted and the received beam should be used. Due to the thin light beams the overlap of the emitted and received beam from a ceilometer with two parallel optical axes can be insufficient in this height range. Further on, Doppler shifts are not analyzed by ceilometers. Therefore, in contrast to acoustic remote sensing with Doppler-SODARs, additional variables in addition to the backscatter intensity are not available from ceilometers for the design of determination schemes for

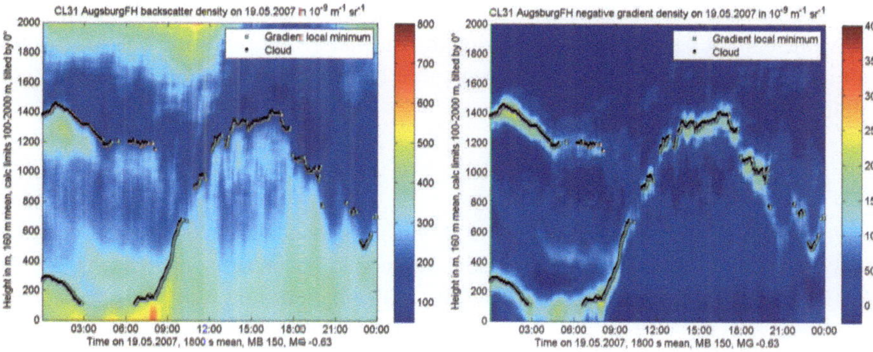

Fig. 4.3 Detection of MLH from ceilometer measurements in Augsburg, Germany on May 19, 2007 (time is given in GMT+2). *Left*: optical backscatter intensity, *right*: negative vertical gradient of optical backscatter intensity. *Dots* indicate MLH determined from the gradient method

MLH. Thus the schemes listed below all resemble to the ARE methods for acoustic remote sensing.

Figure 4.3 shows a sample measurement with a mono-axial ceilometer. The left-hand frame displays the optical backscatter intensity and the right-hand frame the negative vertical derivative of this intensity as time-height sections over one day (from midnight to midnight) and over a height range of 2000 m. The data were received on a clear day in spring, and the vertical structure of the ABL was dominated by surface heating due to incoming solar radiation during daytime and radiative surface cooling during night-time. In the morning hours until about 9 a.m. a shallow stable nocturnal surface layer with a depth of about 200 m and a residual layer with a depth of about 1200–1400 m can be distinguished. From 9 a.m. onwards the evolution of a daytime convective boundary layer with a maximum depth of about 1400 m can be clearly seen. The dots in both frames of Fig. 4.3 indicate the mixing-layer height determined with the gradient method described below (see section "Gradient or Derivative Methods"). The right-hand frame in Fig. 4.3, which shows the vertical derivative of the optical backscatter intensity, demonstrates that the analyzed MLH values indeed coincide with maxima of the negative vertical gradient of this backscatter intensity.

Threshold Method

Melfi et al. (1985) and Boers et al. (1988) used simple signal threshold values, though this method suffers from the need to define them appropriately (Sicard et al. 2006). *H*4 is defined here as the height within the vertical profile of the optical backscatter intensity where the backscatter intensity first exceeds a given threshold when coming downward from the free unpolluted troposphere. The determination of several heights *H*4_*n* would require the definition of several thresholds, which probably cannot be done a priory to the analysis. Therefore this will always lead to a subjective analysis of MLH. The left-hand frame in Fig. 4.3 shows that the threshold value cannot be kept constant during the diurnal evolution of the boundary layer in order to get a result, which is comparable to the one from the gradient method applied in Fig. 4.3.

Gradient or Derivative Methods

Hayden et al. (1997) and Flamant et al. (1997) proposed to use the largest negative peak of the first derivative of the optical attenuated backscatter intensity ($B(z)$) for the detection of $H4$ from LIDAR data (height of gradient minimum $H4_{GM}$),

$$H4_{GM} = \min\left(\frac{\partial B(z)}{\partial z}\right), \tag{4.5}$$

The right-hand frame of Fig. 4.3 demonstrates that this is a very meaningful assumption. Likewise, Wulfmeyer (1999) used the first minimum of the slope to detect the top of a convective boundary layer from DIAL data. Münkel and Räsänen (2004), Münkel (2007), and Schäfer et al. (2004, 2005) applied the gradient method to ceilometer data. Menut et al. (1999) took the minimum of the second derivative of $B(z)$ as the indication for MLH,

$$H4_{IPM} = \min\left(\frac{\partial^2 B(z)}{\partial z^2}\right), \tag{4.6}$$

This method is called inflection point method (IPM). It usually gives slightly lower values for $H4$ than the gradient method (4.5). A further approach was suggested by Senff et al. (1996). They looked for the largest negative gradient in the logarithm of the backscatter intensity (height of logarithmic gradient minimum $H4_{LGM}$),

$$H4_{LGM} = \min\left(\frac{\partial \ln B(z)}{\partial z}\right), \tag{4.7}$$

This approach usually gives the largest value for $H4$. According to Sicard et al. (2006) $H4_{IPM}$ from (4.6) is closest to the MLH derived from radiosonde ascents via the Richardson method. The other two algorithms (4.5) and (4.7) give slightly higher values. The vertical profiles shown in Fig. 4.4 (taken from Emeis et al. 2008) give a comparison of the determination of mixing layer heights from eqs. (4.5) to (4.7).

In Emeis et al. (2007b) the gradient method (4.5) has been further refined and extended to enable the calculation of up to $n = 5$ lifted inversions. This algorithm, which has also been used for the MLH analysis shown in Fig. 4.3, is described in the following. Prior to the determination of gradient minima, the overlap and range corrected attenuated backscatter profiles have to be averaged over time and height to suppress noise-generated artefacts. Therefore, the $H4$ values are determined in a two-step procedure. Between 140 m and 500 m, height sliding averaging

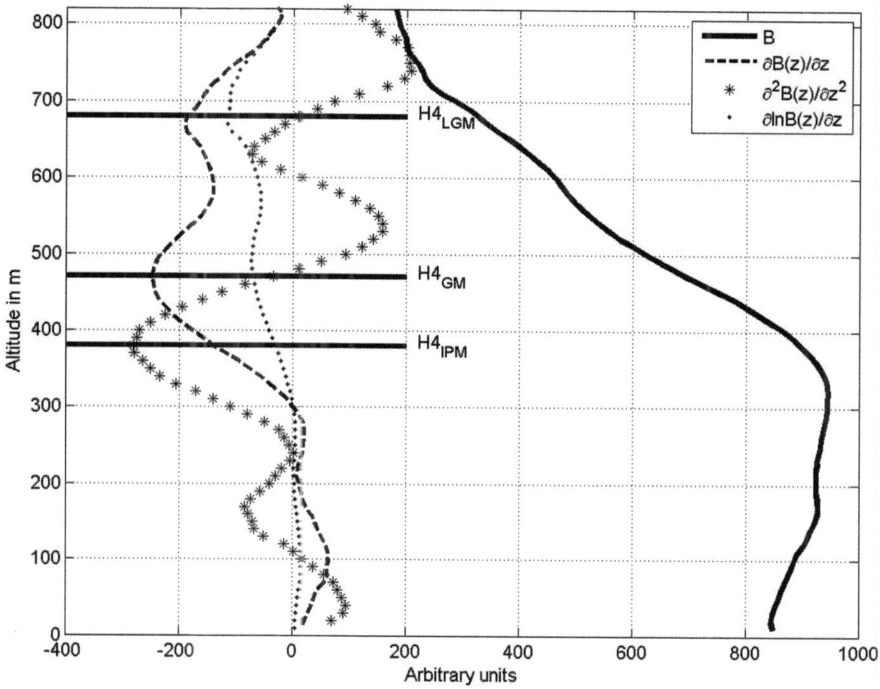

Fig. 4.4 Schematic of the determination of MLH from optical sounding. *Full black curve*: optical backscatter intensity *B* in dB, *dashed line*: vertical derivative of *B*, *asterisks*: vertical second derivative of *B*, *dots*: vertical derivative of the logarithm of *B*. *Horizontal bars* indicate mixing layer heights from eqs. (4.5) to (4.7). From: Emeis et al. (2008)

is done over 15 min and a height interval Δh of 80 m. In the layer between 500 and 2000 m, Δh for vertical averaging is extended to 160 m. Two additional parameters have been introduced to further reduce the number of false hits. The minimum accepted attenuated backscatter intensity $B_{\min}$ right below a lifted inversion is set to $200 \times 10^{-9}\text{m}^{-1}\ \text{sr}^{-1}$ in the lower layer and $250 \times 10^{-9}\text{m}^{-1}\ \text{sr}^{-1}$ in the upper layer. Additionally, the vertical gradient value $\partial B/\partial z_{\max}$ of a lifted inversion must be more negative than $-0.30 \times 10^{-9}\text{m}^{-2}\ \text{sr}^{-1}$ in the lower layer and more negative than $-0.60 \times 10^{-9}\text{m}^{-2}\ \text{sr}^{-1}$ in the upper layer.

If $B(z)$ denotes the measured attenuated backscatter intensity in the height z above ground averaged over time and height and Δh is the height averaging interval, then the gradient $\partial B/\partial z$ in the height z is calculated as

$$\left.\frac{\partial B}{\partial z}\right|_z = \left(B(z + \frac{\Delta h}{2}) - B(z - \frac{\Delta h}{2})\right) / \Delta h, \tag{4.8}$$

A gradient minimum is characterized by a change of sign from minus to plus of the second derivative of $B(z)$. The height interval under examination is searched from bottom to top for these gradient minima *H4_n*.

The second derivative of $B(z)$ in the height z is

$$\left.\frac{\partial^2 B}{\partial z^2}\right|_z = \left(\left.\frac{\partial B}{\partial z}\right|_{z+\Delta h/2} - \left.\frac{\partial B}{\partial z}\right|_{z-\Delta h/2} \right) / \Delta h, \tag{4.9}$$

There is a gradient minimum *H4_n* in the height z if the second derivative of $B(z)$ one range gate below z is not positive, if the second derivative of $B(z)$ in the height z is positive, and if the false hit conditions mentioned above are fulfilled:

$$H4 = z, \text{if } \left.\frac{\partial^2 B}{\partial z^2}\right|_{z-1} \leq 0 \text{ and } \left.\frac{\partial^2 B}{\partial z^2}\right|_z > 0 \text{ and } B\left(z - \frac{\Delta h}{2}\right) \geq B_{\text{min}} \text{ and } \left.\frac{\partial B}{\partial z}\right|_z \leq \frac{\partial B}{\partial z}_{\text{max}}$$
$$\text{for n} = 1, \ldots 5. \tag{4.10}$$

The MLH from optical remote sensing is taken as the lowest height *H4_n*:

$$\text{MLH}_{\text{op}} = H4\_1, \tag{4.11}$$

Idealized Backscatter Method

A parallel development by Eresmaa et al. (2006) using an idealized backscatter profile, originally described by Steyn et al. (1999), is also an extension of the gradient method. MLH is not determined from the observed backscatter profile, but from an idealized backscatter profile fitted to the observed profile. The robustness of this technique is founded on utilizing the whole backscatter profile rather than just the portion surrounding the top of the mixing layer. In this method an idealized backscattering profile $B_i(z)$ is fitted to measured profile by the formula

$$B_i(z) = \frac{B_m + B_u}{2} - \frac{B_m - B_u}{2} erf\left(\frac{z-h}{\Delta h}\right), \tag{4.12}$$

where B_{m} is the mean mixing layer backscatter, B_{u} is the mean backscatter in air above the mixing layer, and Δh is related to the thickness of the entrainment layer capping the ABL in convective conditions. Two new parameters $A1$ and $A2$ are defined so that $A1 = (B_{\text{m}} + B_{\text{u}})/2$ and $A2 = (B_{\text{m}} - B_{\text{u}})/2$. The value of $A1$ is kept constant during the fitting procedure. A good estimation of $A1$ based on an initial order-of-magnitude guess for the MLH is crucial for the quality of the result.

Wavelet Method

A wavelet method has been developed for the automatic determination of mixing layer height from backscatter profiles of an LD-40 ceilometer by de Haij et al. (2006). Before that wavelet transforms have been applied in recent studies for MLH determination from LIDAR observations (e.g. Cohn and Angevine, 2000; Davis et al. 2000; Brooks, 2003; Wulfmeyer and Janjić 2005). The most important advantage of wavelet methods is the decomposition of the signal in both altitude as well as vertical spatial scale of the structures in the backscatter signal.

The wavelet algorithm in de Haij et al. (2006) is applied to the 10 min averaged range and overlap corrected backscatter profile $B(z)$ within a vertical domain of 90–3000 m. For each averaged profile the top of two significant aerosol layers are detected in order to detect MLH as well as the top of a secondary aerosol layer, like, e.g. an advected aerosol layer or the residual layer. This wavelet MLH method uses the scale averaged power spectrum profile $W_B(z)$ of the wavelet transform with 24 dilations between 15 and 360 m and step size 15 m. The top of the first layer, *H4_1*, is detected at the first range gate at which the scale averaged power spectrum $W_B(z)$ shows a local maximum, exceeding a threshold value of 0.1. This threshold value is empirically chosen, based on the analysis of several cases with both well pronounced and less clearly pronounced mixing layer tops. *H4_2* is optionally determined in the height range between *H4_1* and the upper boundary of detection. A valid *H4_2* is detected at the level with the strongest local maximum of $W_B(z)$ provided that this maximum is larger than the $W_B(z)$ of *H4_1*. MLH is set equal to *H4_1*.

However, problems with this method arise e.g. in case of multiple (well defined) aerosol layers, which renders the selection of the correct mixing layer top ambiguous. Furthermore, in spring and summer the detection of the MLH for deep (convective) boundary layers often fails. This is mostly due to the high variability of the aerosol backscatter signal (see Section 4.2.2.5) with height which limits the range for MLH estimation in those conditions (de Haij et al. 2006).

Variance Method

At the top of the convective boundary layer (CBL) we have entrainment of clear air masses from the free troposphere into the ABL. The entrainment process is temporarily variable and leads locally to considerable fluctuations in the aerosol concentration. Therefore, the maximum in the vertical profile of the variance of the optical backscatter intensity can be an indicator for an entrainment layer on top a CBL (Hooper and Eloranta 1986; Piironen and Eloranta 1995). The method is called variance centroid method in Menut et al. (1999). The variance method for the CBL height is also described in Lammert and Bösenberg (2006). Due to the assumptions made this method is suitable for daytime convective boundary layers only. An elucidating comparison between the gradient method and the variance method can be found in Martucci et al. (2004) although they used a Nd:YAG LIDAR at 532 nm instead of a ceilometer and thus suffered from a high lowest range gate in the order of 300 m.

4.2.1.3 RASS

The acoustic and optical methods for the determination of the mixing height, which have been described so far, are all indirect methods that try to infer the mixing height from other variables which usually adapt to the vertical structure of the

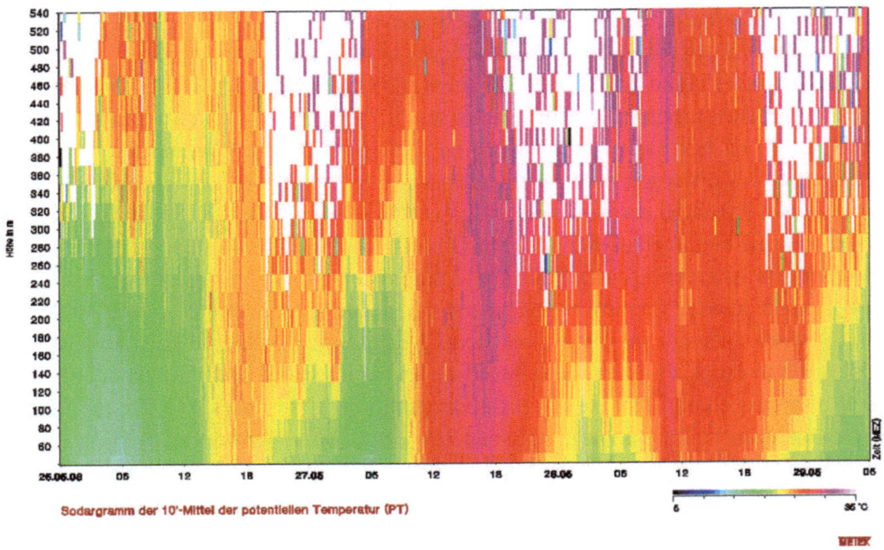

Fig. 4.5 Sample plot of a time-height cross section of potential temperature from RASS measurements over Augsburg, Germany, for a height interval between 40 and 540 m and over 72 hours from May 26 to 28, 2008. *Green*: between 15 and 20°C, *red*: between 25 and 30°C. Temporal resolution is 10 min, vertical resolution is 20 m

ABL. The only direct and key variable for the analysis of the presence of a mixing layer is the vertical profile of virtual temperature. Temperature profiles can directly be measured with a radio-acoustic sounding system (RASS). Figure 4.5 shows an example. There is also the option to derive vertical temperature profiles from Raman-LIDAR soundings (Cooney 1972) and passive radiometer measurements, but especially from passive remote sensing the vertical resolution is usually not sufficient for boundary-layer research. See Section 4.3.2 below for further details.

MLH can be determined from the lowest height where the vertical profile of potential temperature increases with height, indicating stable thermal stratification of the air. The great advantage of RASS measurements is that the magnitude of stability (inversion strength) can be assessed quantitatively, which is not possible from the acoustic and optical sounding devices described before. Figure 4.5 displays a time-height cross section of potential temperature over three days starting at midnight for a height range of 540 m. On the afternoons of the second and the third day well-mixed boundary layers formed, which can be deduced from the vertically constant potential temperature. New surface layers form on the evenings of all three days at about 6 p.m. The depths of these new surface layers increase during the night to about 200 to 300 m. Above these nocturnal surface layers low-level jets (see below) form, indirectly visible from the white areas indicating missing data in Fig. 4.5. Stronger winds like those in low-level jets blow the sound pulses from the RASS out of the focus of the electro-magnetic antenna and hence lead to a failure of the temperature detection.

Ideally, thermal stratification of air should be analyzed from the virtual potential temperature ($\theta_v = \theta\,(1 + 0.609\ q)$), where q is specific humidity) in order to include the effects of the vertical moisture distribution on the atmospheric stability. Unfortunately, no active remote-sensing device for the determination of high-resolution moisture profiles is available. Therefore, the acoustic potential temperature ($\theta_a = \theta\,(1 + 0.513\ q)$), which actually is the temperature that is delivered by a RASS, is often used as a substitute. This is sufficient for cold and dry environments, but somewhat underestimates the virtual potential temperature in humid and warm environments. In case of larger vertical moisture gradients and small vertical temperature gradients, this can lead to a switch in stability from stable to unstable (especially so in marine boundary layers where relative humidity typically decreases with height (see, e.g., Edson et al. 2004)) or vice versa. The following two subchapters give two examples where RASS has been used for MLH determination.

Combined Deployment of Two Different RASS

Engelbart and Bange (2002) have analyzed the possible advantages of the deployment of two RASS instruments, a SODAR-RASS (i.e. a SODAR with an electromagnetic extension) and a high-UHF WPR-RASS (i.e. a wind profiler with an additional sound source), to derive boundary layer parameters. With these instruments, in principle, MLH can either be determined from the temperature profiles or from the electromagnetic backscatter intensity. The latter depends on temperature and moisture fluctuations in the atmosphere. The derivation of MLH from the temperature profile requires a good vertical resolution of the profile, which is mainly available only from the SODAR-RASS. But even if the inversion layer at the top of the boundary layer is thick enough, due to the high attenuation of sound waves in the atmosphere, also the 1290 MHz-WPR-RASS used by Engelbart and Bange (2002) can measure the temperature profile only up to about 1 km. Therefore, in the case of a deeper CBL, MLH was determined from a secondary maximum of the electromagnetic backscatter intensity, which marks the occurrence of the entrainment zone at the CBL top. Thus, with this instrument combination the whole diurnal cycle of MHL is ideally monitored by interpreting the temperature profile from the SODAR-RASS at night-time and by analyzing the electromagnetic backscatter intensity profile from the WPR-RASS during daytime.

Further Algorithms Using a RASS

Hennemuth and Kirtzel (2008) have recently developed a method that uses data from a SODAR-RASS and surface heat flux data. MLH is primarily detected from the acoustic backscatter intensity received by the SODAR part of the SODAR-RASS and verified from the temperature profile obtained from the RASS part of the instrument. Surface heat flux data and statistical evaluations complement this rather complicated scheme. The surface heat flux is used to identify situations with unstable stratification. In this respect this observable takes over an analogous role as the σ_w in the EARE algorithm (see section "Enhanced Acoustic Received Echo Method"). The results have been tested against radiosonde soundings. The coincidence was good in most cases, except for a very low MLH at or even below the first range gate of the SODAR and the RASS.

4.2.1.4 Other Algorithms Using More Than One Instrument

Using more than one instrument for sounding can help to overcome some of the above described deficiencies (limited vertical range, limited data availability) of the individual instruments. Possible combinations are listed in the following subsections.

Combined Deployment of SODAR and Wind Profiler

Beyrich and Görsdorf (1995) have reported on the simultaneous usage of a SODAR and a wind profiler for the determination of MLH. For the SODAR data the ARE method was used. From the wind profiler data MLH was likewise determined from the height of the elevated signal intensity maximum (see also Angevine et al. 1994; Grimsdell and Angevine 1998, White et al. 1999). Good agreement between both algorithms was found for evolving convective boundary layers. The vertical ranges of the two instruments (50–800 m for the SODAR and 200–3000 m for the wind profiler) allowed following the complete diurnal cycle of MLH.

Combined Deployment of SODAR and Ceilometer

There is an interesting difference between the schemes for the determination of MLH from acoustic and optical backscatter intensities, which should be noted carefully. While the acoustic backscatter intensity itself is taken for the detection of $H1$ and $H3$ (see (4.1) and (4.3)) and the first derivative of this backscatter intensity for the determination of $H2$ (see (4.2)), the first and the second derivative of the optical backscatter intensity (but not the optical backscatter intensity itself) is used to determine $H4$ (see (4.5)). This discrepancy in the processing of the two backscatter intensities is due to the different scattering processes for acoustic and optical waves: Acoustic waves are scattered at atmospheric refractivity gradients and thus at temperature gradients (Neff and Coulter 1986) while optical waves are scattered at small particles. Therefore, the optical backscatter intensity is proportional to the aerosol concentration itself. The MLH, on the other hand, which we desire to derive from these backscatter intensities, is in both cases found in heights where we have vertical gradients of the temperature and of the aerosol concentration. Therefore, in principle, the vertical distribution of the acoustic backscatter intensity should look very much alike to the negative of the vertical distribution of the vertical gradient of the optical backscatter intensity. The schematic diagrams in Fig. 4.6 are illustrating this fact.

Simultaneous measurements with different remote-sensing devices have mainly been made in order to evaluate one remote-sensing method against the other (Devara et al. 1995). But one could also think of combining the results two or more remote-sensing devices for determining the structure of the ABL. Direct detection of MLH from acoustic backscatter intensities is limited to the order of about 1 km due to the

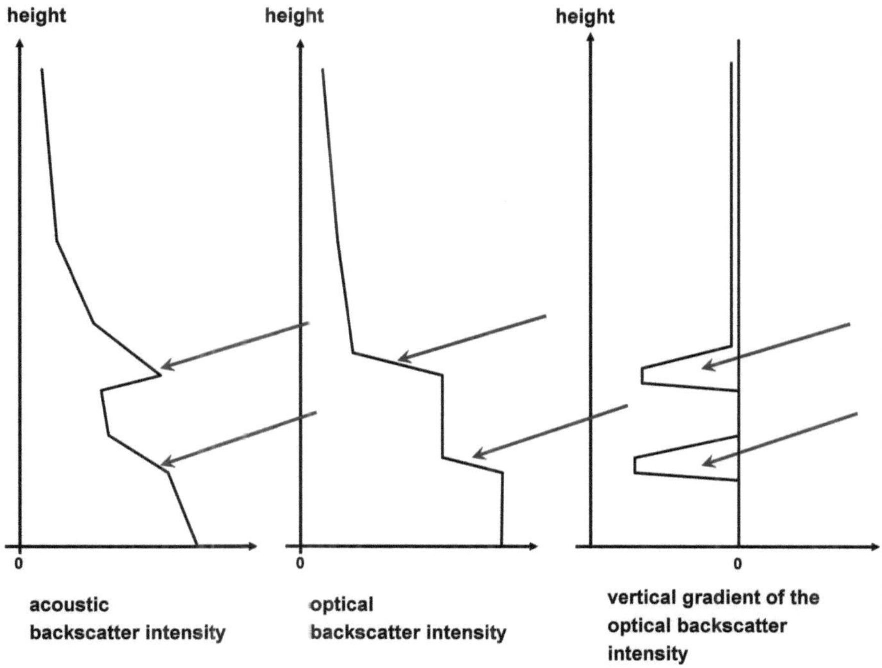

Fig. 4.6 Comparison of schematics for the MLH determination from SODAR (left-hand graph) and ceilometer (middle and right graphs). The right-hand graph shows the vertical derivative of the middle graph. Therefore, the left-hand graph and the negative of the right-hand graph have many similarities

rather high attenuation of sound waves in the atmosphere. In contrast, optical remote sensing offers much larger height ranges of at least several kilometres, because the attenuation of light waves in the atmosphere is small unless there is fog, clouds, or heavy precipitation. The analysis of the SODAR data and the ceilometer data can be combined to one single piece of MLH information by forming the minimum from (4.4) and (4.11):

$$\mathrm{MLH} = \min(\mathrm{MLH_{ac}}, \mathrm{MLH_{op}}), \tag{4.13}$$

4.2.1.5 Correlation of Pollutant Concentrations with MLH

This subsection is to give examples for the application of continuous mixing layer depth information from ground-based remote sensing.

Concentration of pollutants which are emitted at or formed near the surface should correlate with MLH, because the depth of the mixing layer defines the size of the atmospheric volume in which these pollutants can dilute. This correlation should be best for aged inert substances away from their direct sources that have already efficiently been mixed in the vertical. For the investigation of these correlations, the influence of horizontal advection should be negligible, either due

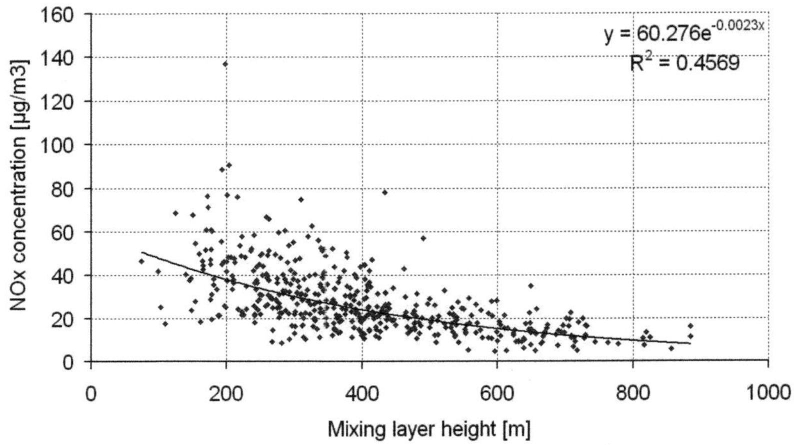

Fig. 4.7 Negative correlation between SODAR-derived mixing layer height and NO_x concentrations measured at an urban background station. From Schäfer et al. (2006a)

to low winds or due to vanishing horizontal concentration gradients. Schäfer et al. (2006a) give examples of such correlations between MLH, derived from SODAR measurements, and NO_x-concentrations and aerosol concentrations observed at an urban background station. Figure 4.7 shows the considerable negative correlation between MLH and NO_x-concentrations. An increase in the mixing layer height by a factor of four from 200 to 800 m leads to a reduction of the NO_x concentration by a likewise factor of four from 40 to 10 $\mu g/m^3$. Thus, in this data range, the correlation between pollutant concentration and mixing layer height can be perfectly described by a hyperbola.

If the above preconditions apply, the simultaneous measurements of pollutant or trace gas concentrations and MLH could be used to infer the emission rate of these substances. For a given constant MLH and perfect vertical mixing, the concentration should increase with a rate inversely proportional to MLH. Such pollutant or trace gas emission rates are then representative for areas of the order of 100 km^2, if MLH is horizontally homogeneous over this area, if horizontal winds are of the order of a few m/s, and if vertical mixing takes place within about one hour. This budget or mass balance method for emission determination (see Denmead 2008 for an overview of emission determination techniques) is especially suited for the analysis of emission rates of diffusely emitted substances where it is difficult to localize specific emission sites. An example for the derivation of the regional emission rate of methane from a rural area from MLH, derived from SODAR measurements, and near-surface nocturnal methane concentration increases is shown in Emeis (2008). Figure 4.8 is taken from this study showing a time series of the near-surface methane concentration (path-averaged values from FTIR measurements over several hundreds of metres), a simultaneous time series of MLH from SODAR measurements, and the deduced nocturnal emission rates (indicated by the hyperbolae in the lower frame of Fig. 4.8) from the nocturnal methane concentration increases. This is the

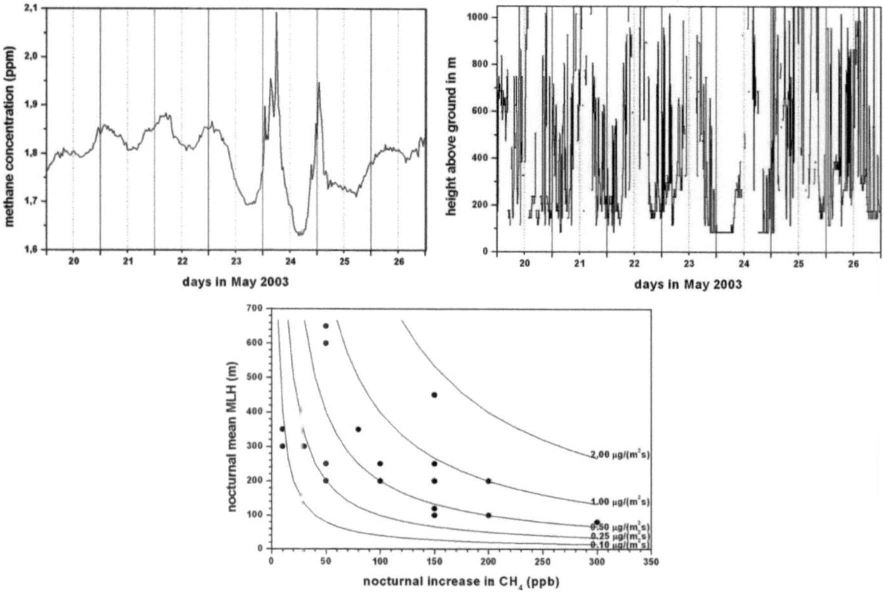

Fig. 4.8 Methane emission rates (below, for 18 nights in May 2003) in µg/(m^2 s^{-1}) from path-averaged FTIR methane concentration measurements (*upper left*, time series for seven consecutive days in May 2003) in ppm and mixing layer heights derived from SODAR measurements (*upper right*, same time period as for the methane concentrations in the upper left frame) in m

same type of hyperbolic behaviour that we have also noticed in Fig. 4.7. Comparison of the methane and the MLH time series clearly indicates that the largest methane concentration increases happen during the nights with the lowest MLH. The order of magnitude of the deduced areal emission rates fits to the nationwide German methane emission rates as documented in the Kyoto protocol compliance reports.

4.2.2 Boundary Layer Height

Often, the boundary layer consists of more layers than just the mixing layer. For example, at night, a residual layer may persist over a newly formed near-surface stable surface layer. The combined deployment of a SODAR and a ceilometer (see section "Combined Deployment of SODAR and Ceilometer") may be a choice to detect such features. Such a combination of parallel measurements of the vertical structure of the atmospheric boundary layer by a ceilometer and a SODAR is described in Emeis and Schäfer (2006). Figure 4.9, which is taken from this study, shows a daytime convective boundary layer, shallow nocturnal surface layers in the morning and the evening, and a residual layer above the nocturnal surface layers. The ceilometer detects the overall boundary layer height (blue triangles) whose height is partly modified by large-scale sinking motion in the anticyclone dominating the weather during the measurement period. Stable nocturnal surface layers

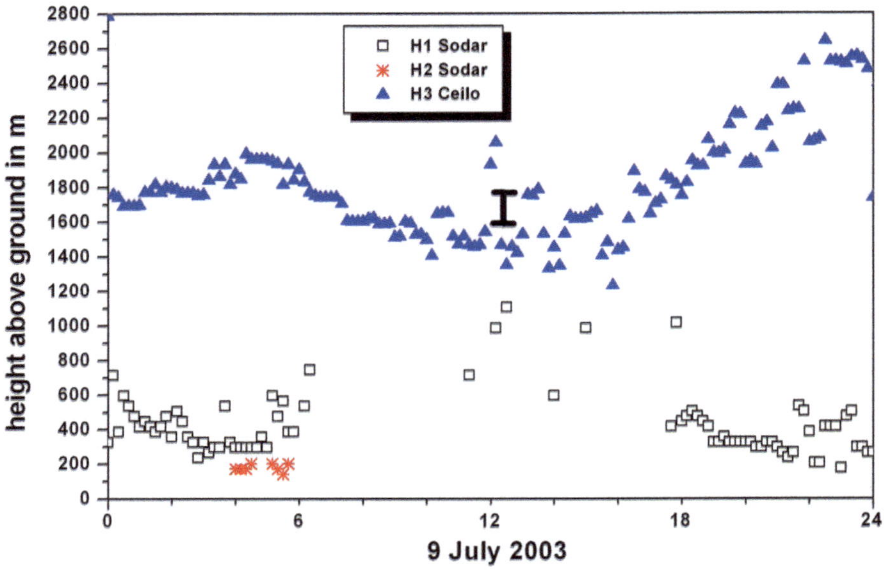

Fig. 4.9 Boundary layer height and layering from combined ceilometer and SODAR measurements over Budapest, Hungary, on July 9, 2003. *Black squares* are from eq. (4.1), *red asterisks* are from eq. (4.2) and (4.3), *blue triangles* are from eq. (4.10) (From Emeis and Schäfer 2006)

with a depth of a few hundred metres can be detected underneath the black squares derived from the SODAR soundings. The convective boundary layer during daytime fills the full depth of the boundary layer. This combination offers the same advantages as the combination of SODAR and wind profiler presented in Beyrich and Görsdorf (1995). First results from a combined deployment of a RASS and a ceilometer are given in Emeis et al. (2009).

In such combined remote sensing measurements a SODAR better detects the near-surface features such as nocturnal stable layers (a RASS instead of a SODAR directly delivers the near-surface temperature profile) while the ceilometer is able to follow the diurnal variation of the daytime convective boundary layer and the top of the whole boundary layer. The residual layer then becomes visible as the layer between the new nocturnal surface layer below and the top of the boundary layer above.

4.2.3 Clouds and Cloud Base Height

The detection of the base height of clouds at the top of the atmospheric boundary layer is one of the simplest remote-sensing tasks to be executed. The oldest method, a cloud searchlight, does not involve sounding but is based only on geometrical calculations. For this bistatic method a strong searchlight produces a light fleck on the lower edge of the cloud by a vertically directed light beam. An observer, who is located in a known distance from the position of the searchlight, measures the angle

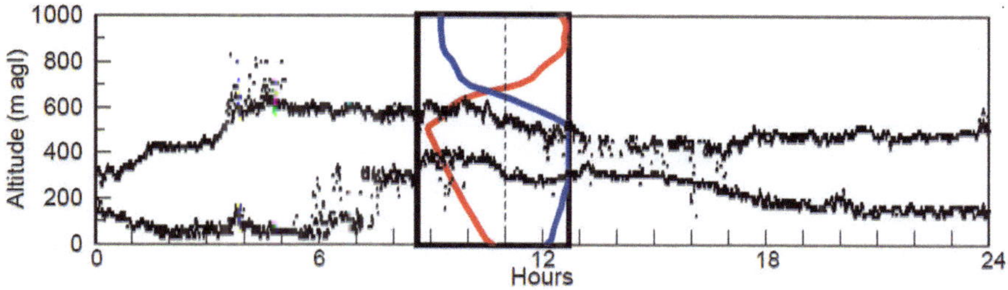

Fig. 4.10 Time series of cloud base (estimated from ceilometer) and cloud top (estimated from cloud radar), December 9, 2003. The *rectangle* in the middle of the figure corresponds to the profiles measured with a radiosounding at 11 UTC (*dashed line*): *red line* = temperature with a horizontal scale of 10 K, and *blue line* = humidity with a horizontal scale of 100%. (Fig. 6 from Ruffieux et al. 2006)

under which the light fleck is seen and then computes the cloud base height from simple trigonometric formulas.

Today, following the increasing pressure to automate observations, mono-static methods are used which are based on the measurement of time delay. This is the original field of operation for ceilometers, which nowadays are also used for determining aerosol profiles in the boundary layer (see below).

Fog and low cloud top and base heights may be derived from 78 GHz frequency-modulated continuous wave (FMCW) cloud RADAR and laser ceilometer measurements. The base layer is well measured with a ceilometer while the top of the cloud is determined using the 78 GHz FMCW RADAR. This method for cloud top determination is based on determining the height at which the RADAR signal falls below a defined threshold level. This W-band RADAR is most sensitive to drizzle size drops, so it is not sensitive to clouds with only very small drops. Thus, in thin mist the cloud RADAR may not provide an indication of the fog top. The rapid decrease of the RADAR sensitivity below 120 m above ground level is another limitation in the measurement of low fog layers (Ruffieux et al. 2006). Figure 4.10 shows an example for a simultaneous cloud top detection by a cloud RADAR and cloud base detection by a ceilometer, and validates the results by a comparison to radiosonde measurements. There was fog at night of this winter day, as the cloud base was nearly at the ground before 6 a.m. Generally, cross sections through non-precipitating clouds can be obtained from Ka- and W-band RADARs, which operate at wavelengths of a few millimetres (see also Table 3.2).

4.3 Vertical Profiles of Atmospheric Variables

So far, we have dealt with the general vertical structure of the ABL and the demarcation of several sublayers of the ABL. For many purposes, detailed profiles of certain atmospheric variables such as wind speed, temperature, or humidity must be known throughout the ABL with reasonable accuracy and high vertical resolution (see Table 4.2).

Table 4.2 Overview of methods to derive vertical profiles of atmospheric variables using ground-based remote sensing (see rightmost column for section number)

Variable	Detection method	Section
Wind speed, direction, turbulence	SODAR	4.3.1.1
Wind speed, direction, turbulence	Wind LIDAR	4.3.1.2
Wind speed, direction, turbulence	RADAR windprofiler	4.3.1.3
Temperature	RASS	4.3.2.1
Temperature	Raman LIDAR	4.3.2.2
Temperature	Radiometer	4.3.2.3
Temperature	Interferometer	4.3.2.4
Humidity	DIAL	4.3.3.1
Humidity	Raman LIDAR	4.3.3.2
Humidity	Radiometer	4.3.3.3
Humidity	Interferometer	4.3.3.4
Rain	Micro rain RADAR	4.3.3.5
Trace gases	Vertical profiling	4.3.4.1
Trace gases	Horizontal path-averaging	4.3.4.2

4.3.1 Wind and Turbulence

Initially, remote sensing of the wind in the atmospheric boundary layer was mainly done with acoustic methods (SODAR), because the speed of sound which is only one to two orders of magnitude larger than the wind speed leads to relatively large Doppler shifts that are easily detectable (Neff and Coulter 1986). Additionally, this method offers a high vertical resolution (5–10 m) and a small lowest range gate (10–30 m), which makes it suitable for investigating wind profiles in shallower boundary layers as well. The technical and electronic obstacles to overcome for the application of the acoustic method are considerably lower and cheaper than for the respective optical and electromagnetic methods. The advantages are neutralized by the limited data availability of the acoustic method. In fact, none of the remote-sensing methods (SODAR, RADAR windprofiler, wind-LIDAR) can guarantee a one hundred percent data availability for a given height due to the backscattering properties of the atmosphere. The strongest limitation by atmospheric conditions with respect to wind measurements is found with the weather RADAR technique that delivers reasonable data only in case of rain, although under limited conditions signals from insects or index-of-refraction turbulence have been used in clear-air boundary-layer studies. A SODAR fails in situations with missing turbulence and low temperature gradients. These situations lead quite often to a low range of a SODAR around dawn and dusk (Maughan et al. 1982; Emeis et al. 2004). A RADAR windprofiler usually suffers from very dry air masses only, and the LIDAR technique is hampered by strong rain, fog, and clouds.

If all three components of the wind have to be measured by sounding methods, then successive velocity-component profiles have to be made in at least three different directions shortly after each other. The wind components can subsequently be computed from the radial velocities by applying trigonometric relations. SODARs and windprofilers perform these three-dimensional wind soundings by emitting upward three to five beams inclined to each other by about 15° to 20° (Doppler-beam-swinging method). RADARs and LIDARs usually make conical scans around the zenith with comparable aperture angles. This scanning mode is also called VAD (velocity-azimuth-display) mode (Browning and Wexler 1968; Banta et al. 2002). For calculation of a mean wind, it has to be assumed that the wind field is homogeneous over the horizontal distance enclosed by the beams or the conical scan. Thus, remote-sensing methods due to their design yield values that are representative of larger air volumes. This fact hampers the direct intercomparison with cup anemometer data in locations where the wind flow is non-uniform over the region of space encompassed by the mast and LIDAR scan; but, on the other hand, remote-sensing methods are very likely more suited for wind energy applications because the air volume scanned by the measurement device and the air volume which passes through the rotor plane of a wind turbine are of comparable size.

4.3.1.1 Acoustic Techniques

The operation of Doppler-SODAR devices in the Doppler-beam-swinging (DBS) mode was the first remote-sensing technique to continuously observe vertical profiles of wind speed and direction in the ABL. With these instruments, the increase of wind speed with height and the turning of the wind direction with height over different types of land use (e.g. over large cities) can be investigated. Likewise, wind maxima at inversions and fronts can be observed. Also the modification of the wind profile by the presence of hills, escarpments, and land use changes can be studied.

Diurnal wind phenomena related to topographic features (sea shores, valleys, mountains) and nocturnal phenomena such as low-level jets are dealt with further below in Section 4.5.

Wind and Turbulence Profiles over Flat Terrain

Over flat terrain the vertical wind profile most likely follows the known theoretical wind profile laws given in Chapter 2. The left frame of Fig. 4.11 shows an example, where monthly mean vertical profiles of the scale parameter of the Weibull distribution (the two-parameter Weibull distribution is frequently used to describe the frequency distribution of wind speeds, its scale parameter is proportional to the mean wind speed, while its shape parameter defines the shape of the distribution) over flat terrain are compared to analytical profiles for the Prandtl layer (see (2.5a)) and the wind resource model WAsP (Troen and Petersen 1989). In the example displayed in the left frame of Fig. 4.11, the measured wind profiles follow the analytical profiles quite well for heights up to 60–80 m. This can be taken as an indication that the average depth of the Prandtl layer was of this magnitude. Modifications of such profiles mainly occur by land-use changes and by the passage of synoptic disturbances (fronts).

It was already noticed by the very first researchers who looked into wind profile data (see Section1.1) in 1915 and 1921 that the diurnal variation of wind speed

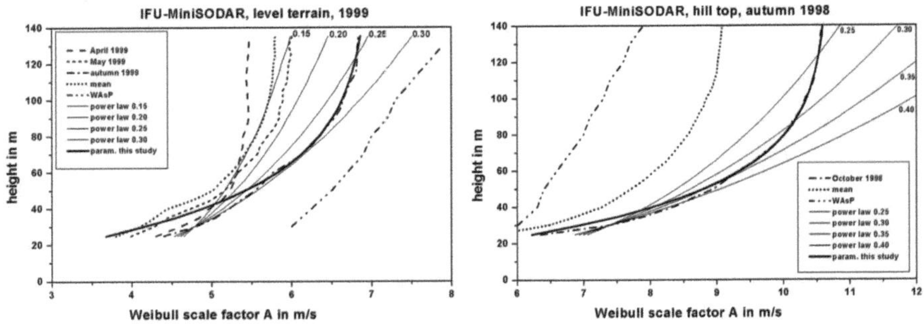

Fig. 4.11 Vertical profiles of the Weibull scale factor in m s^{-1} (which is proportional to the mean wind speed) for flat terrain (*left*) and for a hill top (*right*). Here, SODAR measurements are compared to power law profiles with different exponents (*thin full lines*), the wind resource assessment model WAsP (*dashed double-dotted line*, Troen and Petersen 1989) and a parameterization (*full bold line*) suggested in Emeis (2001)

varies with height on clear days. It is only for the lower heights below about 60–80 m above ground that we have low wind speeds at night and slightly higher wind speeds during daytime. Above 60–80 m above ground this behaviour is reversed: there are lower wind speeds during daytime and higher wind speeds at night. This behaviour is a consequence of the diurnal variation of the thermal stratification and the vertical turbulent mixing during cloudless days, on which radiative fluxes are dominating. Solar radiation is heating the ground during the day and thermal outgoing radiation is cooling it again during the night, leading to stable thermal stratification at night and neutral to unstable stratification during the day. In both cases, the changes around sun rise and sun set start near the ground and then propagate into higher layers of the boundary layer. This explains why the wind speed in the lowest height (see Fig. 4.12) increases shortly after sun rise before decreasing again later in the morning. In these hours the unstable well-mixed surface layer is only about 100–150 m deep and the two wind speeds at 40 and 100 m mix to a nearly uniform value. Later when the well-mixed convective surface layer grows deeper, the wind speed in all heights is decelerated by the surface friction and is only slowly increasing during the day due to the existing large-scale pressure gradient. Around sun set a new stable surface layer forms near the ground. The layers above nearly immediately loose contact to the frictional forces at the ground and speed up. This is the very mechanism which also leads to the formation of the nocturnal low-level jets (see Section 4.5.1).

A likewise diurnal variation can be observed for the atmospheric turbulence on such days. Figure 4.13 displays the diurnal variation of the standard deviation of the vertical wind component for the same case depicted in Fig. 4.12. Lower turbulence is present during nights in a stable nocturnal boundary layer and higher turbulence during daytime hours in a convective boundary layer. Quite noticeable is the increase of turbulence with height in the convective boundary layer during the day, which follows the behaviour given in eq. (2.35). The slight increase of turbulence with height in the early morning hours is due to the wind shear at the lower side of the

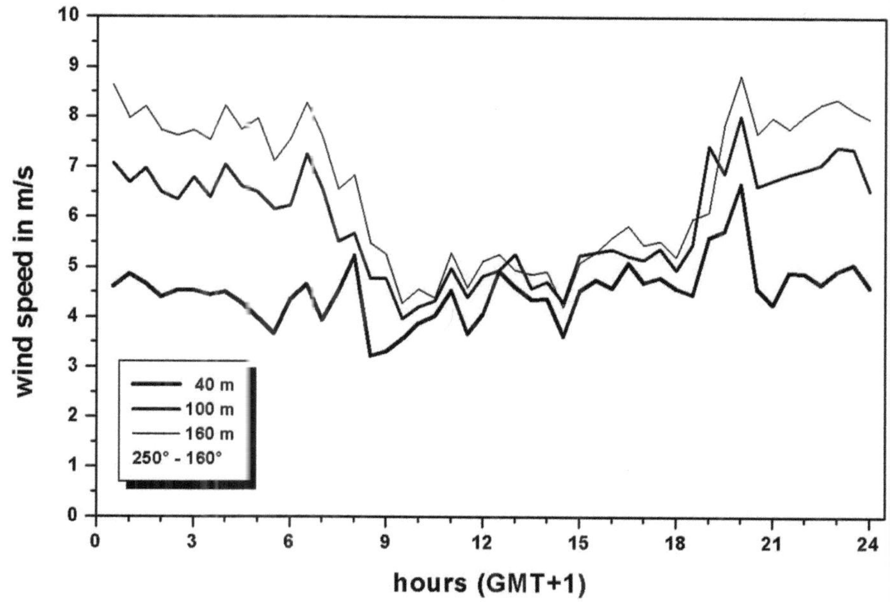

Fig. 4.12 Diurnal variation of the horizontal wind speed in three different heights above ground (40, 100, and 160 m) during an undisturbed cloudless weather situation in summer 2005. The half-hourly data are from SODAR measurements near the Paris airport Charles de Gaulle

low-level jet blowing at night in heights of a few hundred metres above ground (see Section 4.5.1).

Figure 4.14 merges the turbulence (Fig. 4.13) and wind speed (Fig. 4.12) information together in vertical profiles of turbulence intensity. Turbulence intensity is formed here as the ratio of the standard deviation of the vertical wind speed component and the mean horizontal wind speed. This is the only non-dimensional turbulence measure, which can be derived from SODAR measurements. A strong difference between daytime and nocturnal conditions is observed. The daytime turbulence intensity is more than twice the night-time turbulence. The other difference is the vertical slope of the profiles. While the nocturnal profile exhibits a slight decrease with height, the daytime profile shows a clear increase with height. This increase is due to the increase in the standard deviation of the vertical wind speed component, because the mean horizontal wind speed is nearly identical in these three heights during daytime.

Wind and Turbulence Profiles over Cities

Wind and turbulence profiles over cities differ from those over flat rural terrain due to the specific vertical structure of the urban boundary layer (UBL) presented in Chapter 2. Surface-based vertical soundings in the UBL for longer time periods have so far been made only in Moscow (Russia, Pekour and Kallistratova 1993; Pekour

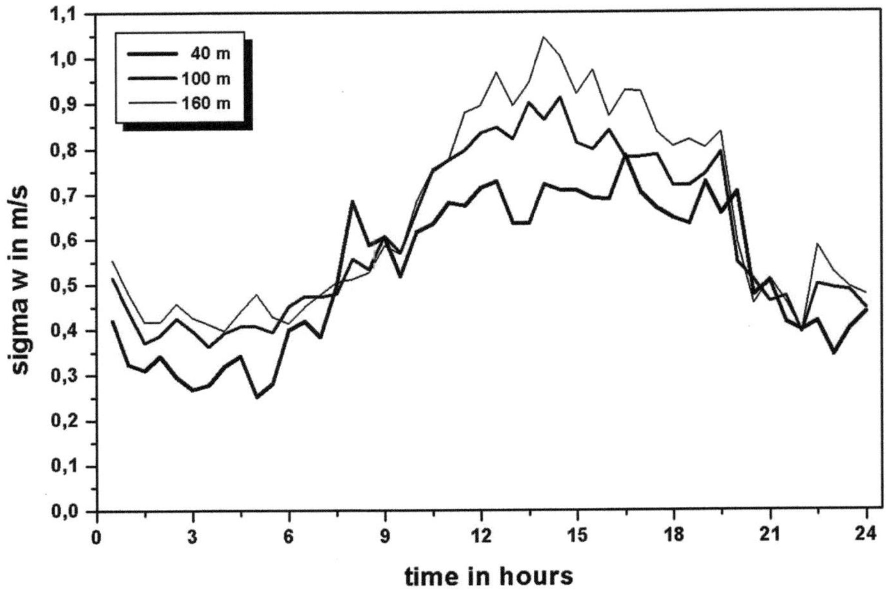

Fig. 4.13 Diurnal variation of the standard deviation of the vertical wind component (sigma w) in three different heights above ground (40, 100, and 160 m) during an undisturbed cloudless weather situation in summer 2005. The half-hourly data are from SODAR measurements near the Paris airport Charles de Gaulle

et al. 1993; Lokoshchenko 2002), in Hannover (Germany, Emeis and Türk 2004), Toronto (Canada) for the CBL (Melling and List 1980), and Linz (Austria, Emeis et al. 2007b). Mean seasonal profiles of wind speed, standard deviation of the vertical velocity and turbulence intensity from SODAR measurements in three cities of different size, Moscow, Hannover, and Linz, have been recently compared to analytical descriptions of the wind profile for the Prandtl and Ekman layer in Emeis et al. (2007b). We will excerpt from this review in the next paragraphs. Comparable studies on wind profiles in the urban boundary layer have been written by Lokoshchenko et al. (2007, 2009) and Kallistratova et al. (2009) as well.

Mean Urban Wind Profiles

Mean monthly wind profiles observed in Hannover in April 2003 together with theoretically derived stability-dependent logarithmic wind profiles are shown as an example in Fig. 4.15. April is representative for spring in Hannover. Nocturnal wind speeds are on average higher than daytime wind speeds except for heights below 125 m above ground, because the atmosphere is stabler at night and less vertical momentum exchange takes place. The monthly mean is computed from 30 min-mean profiles, which are complete up to at least 210 m. Above this height, the data availability decreases and is about 20% at 500 m height.

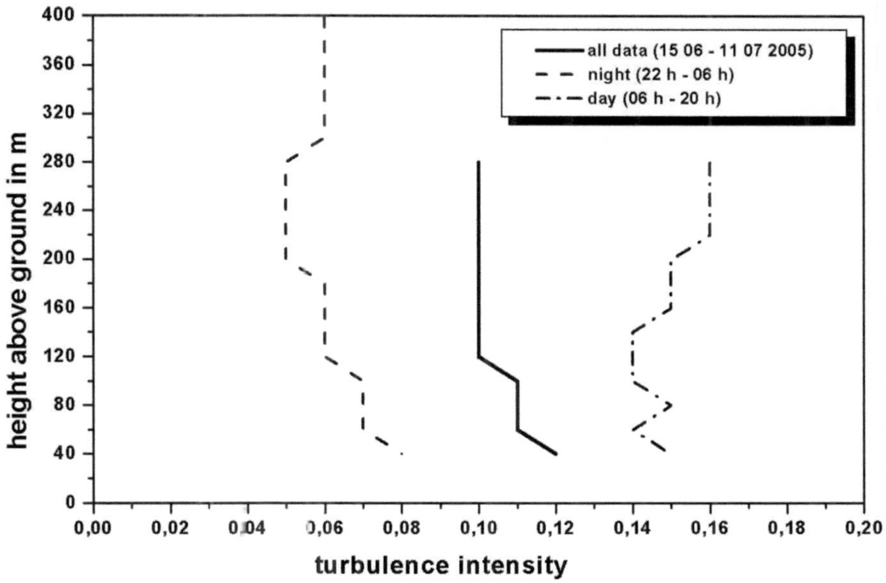

Fig. 4.14 Daytime (*dash-dotted line*), night-time (*dashed*), and daily mean (*full line*) profiles of turbulence intensity from SODAR measurements in summer 2005 near Paris airport Charles de Gaulle

In the fits with the analytical profiles, the roughness length was set to $z_0 = 1$ m. The friction velocity, u_* is varying between 0.356 m s^{-1} for the night-time profile and 0.538 m s^{-1} for the daytime profile. The analytic and the mean observed wind profiles agree up to 350 m above ground, while the observed wind speeds are on average 1–2 m s^{-1} lower than expected according to the logarithmic wind profiles at higher levels.

Mean Urban Turbulence Profiles

The software of SODAR instruments usually delivers the standard deviation of the vertical velocity component, which can be used as a rough measure of atmospheric turbulence. Figure 4.16 shows monthly mean profiles of the vertical component of the turbulence intensity observed in Hannover from SODAR measurements, i.e. the variance of the vertical wind component σ_w divided by the average horizontal wind speed. The daytime values of turbulence intensity in this example are twice as high as the night-time values. At daytime, turbulence intensity profiles are more or less constant with height up to 300–400 m above ground. In autumn, winter, and generally at night-time, the profiles show a strong decrease of the turbulence intensity with height within the lower 150–200 m in accordance with (2.35) together with the first equation of (2.27).

The strong difference to turbulence profiles over rural terrain becomes obvious by comparing Figs. 4.14 and 4.16. While the turbulence measurements over the

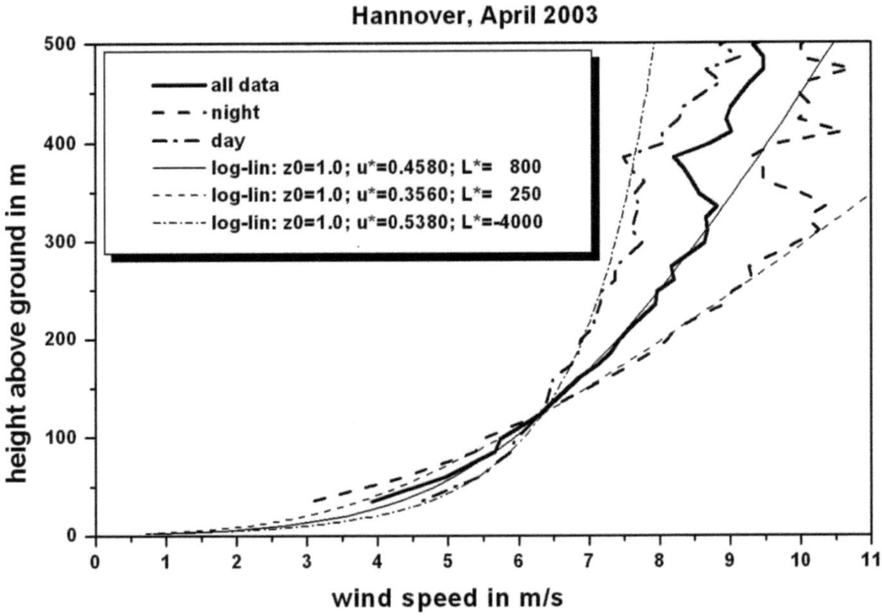

Fig. 4.15 Mean monthly vertical wind profiles (*bold full line*: all data, *dashed line*: nocturnal data, *dashed-dotted* line: daytime data) over a town (Hannover, Germany) for April 2003 from SODAR measurements. The profiles are compared to the diabatic logarithmic wind profile (upper equation in (2.27). Parameters for (2.27) are given in the insert

urban area show constantly high turbulence intensities near the ground in the urban canopy layer, the rural profiles exhibit a strong diurnal variation close to the ground as well. This high turbulence level that is maintained during the night over urban surfaces is mainly due to the large heat storage of the urban surface, which prevents the formation of strong stable stratification in the urban canopy layer during nighttime. This is in agreement with the findings presented by Uno et al. (1988, 1992) and Dupont et al. (1999) who also showed that the large roughness of the town surface and the urban heat island prevents the formation of a cool stably stratified nocturnal surface layer as it is typical for rural areas for nights with clear skies.

Comparison to Analytical Wind Profile Laws

The wind profiles shown in Fig. 4.15 deviate from the analytical profiles from the first equation in (2.27) in greater heights. Therefore, in Fig. 4.17 comparison is made with the analytical curve from the complete equation (2.27). This fit is much better for greater heights. Here the mean monthly nocturnal wind profile observed in Hannover, Germany, in August 2002 is displayed. Apart from a secondary maximum due to the frequent occurrence of low-level jets in this month between 200 and 350 m above ground, the full eq. (2.27) gives a good representation of the wind profile at all heights. As the low-level jet is a non-stationary feature, it is clear that

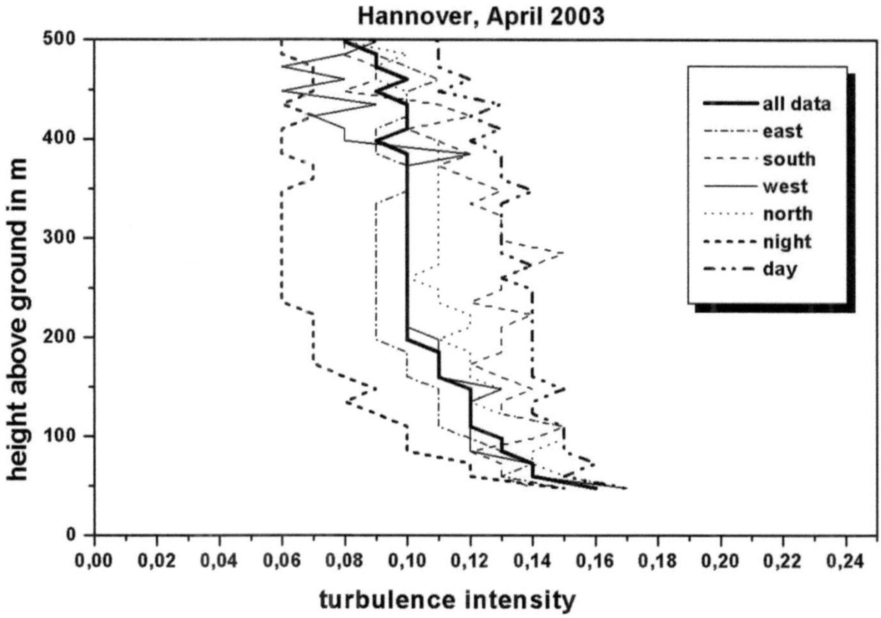

Fig. 4.16 Mean monthly vertical profiles of turbulence intensity over a town (Hannover, Germany) for April 2003 from SODAR measurements (*bold full line*: all data, *bold dashed line*: nocturnal data, *bold dash-dotted line*: daytime data, *thin lines*: data from the four wind direction sectors)

it cannot be described by the stationary eq. (2.27). The three additional free parameters in the second and third equation of (2.27) in comparison to the first equation are tuned in order to improve the agreement to the measured wind profiles. The first three parameters have been kept. For the approximation shown, the Prandtl-layer height, z_p is set to 30 m at night. Below this height both profile laws (first equation and all three equations of (2.27) give identical values. The roughness length, z_0 is set to 1 m, the Monin–Obukhov length, $L*$ is chosen in order to represent the profile below z_p properly, and the geostrophic wind speed is set equal to the wind speed at 500 m. The friction velocity, $u*$ must be iterated as described after eq. (2.16). From this iteration the friction velocity at night-time is $u_* = 0.18 \text{ m s}^{-1}$. The turning angle of the wind direction is around 32° at night.

4.3.1.2 Optical Techniques

The aforementioned acoustic sounding has a limited height range, usually less than one kilometre. Wind LIDARs are an alternative, if data from larger ranges should be obtained. One of the available LIDARs is the scanning High-Resolution Doppler LIDAR (HRDL) recently developed at the Earth System Research Laboratory of the U.S. National Oceanic and Atmospheric Administration (NOAA/ESRL; formerly Environmental Technology Laboratory). HRDL operates at 2.02 μm in the infrared

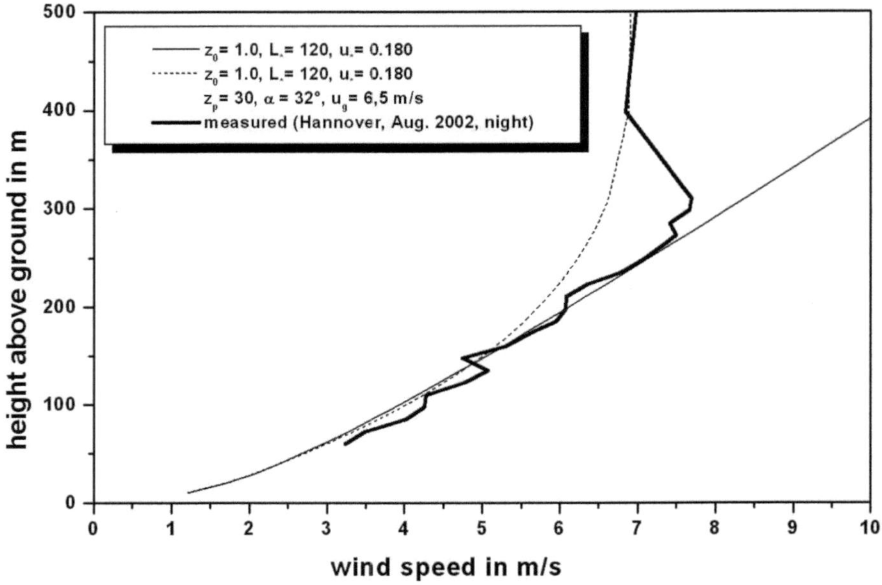

Fig. 4.17 Mean monthly nocturnal wind profile over a town (Hannover, Germany) for August 2002 from SODAR measurements. Also shown are analytical profiles from the first equation in (2.27) (*thin full line*) and from the complete equation (2.27) (*dotted line*)

part of the light spectrum, has a first range gate of data at ~200 m range, a range resolution of 30 m, and a velocity precision evaluated at ~20 cm s^{-1} (Wulfmeyer et al. 2000; Grund et al. 2001). Wind-energy-related measurements were, e.g. taken in the U.S. Great Plains, where a major contribution to the wind resource is a recurrent nocturnal southerly low-level jet (LLJ) that is prevalent during the warm season. Important properties of the LLJ are its speed and height, the speed and shear across the height interval spanned by the rotors, and the magnitude and frequency of turbulence in the rotor layer. These properties were investigated using HRDL at two locations, southeast Kansas during the Cooperative Atmosphere-Surface Exchange Study campaign of October 1999 (CASES-99: Banta et al. 2002; Poulos et al. 2002) and in southeast Colorado, south of the town of Lamar, during the Lamar Low-Level Jet Project (Kelley et al. 2004, 2005; Pichugina et al. 2004, 2005a) in September 2003 (LLLJP-03). In the months immediately following the Lamar deployment, more than 100 wind turbines were installed at the site of LLLJP-03.

Mean wind profiles were determined using azimuth (conical) scans and elevation scans (which produce a vertical slice of data in the atmosphere and which were generally performed parallel to the wind direction) in the abovementioned studies. Frequency distributions of mean wind properties (speed, direction) are useful for site assessment of wind-energy potential (Banta et al. 2002; Pichugina et al. 2004, 2005a). Another property of the wind profile is the power-law exponent often used to extrapolate wind-profile data to altitudes above which measurements were taken. Distributions of this quantity determined by Pichugina et al. (2004) from HRDL data show that the value of the exponent is often much larger than its assumed value of 0.2, and that it varies with height.

Over the U.S. Great Plains, Banta et al. (2002, 2006) found that the directional shear with height was generally small when LLJs of about 12 m s^{-1} or more were present. Under these conditions, along-wind vertical slice scanning can be used to produce wind-speed profiles at intervals of 30 s or less. When significant directional shear is present, the wind profiles would need to be monitored using conical scans at one or more elevation angles, which may take 2–5 min, depending on how many elevations are used for the sampling.

Peña et al. (2009) compare wind profiles well beyond 80 m, obtained with a small wind LIDAR, with their new extended laws (see eqs. (2.21) and (2.33)) for wind profiles in the ABL. In addition to the usefulness of mean-wind data for long-term wind energy site assessment, LIDAR profile data could also be useful for real-time wind turbine operations. Spatially averaged, statistically significant profile data are available at time intervals of 2 min or less, which could be used to adjust turbine operational settings to accommodate changes in wind speed and direction. Although the range resolution of HRDL is 30 m, the desired vertical resolution of ~10 m or less can be achieved by scanning (or pointing) at lower elevation angles, or by performing elevation (vertical-slice) scans.

An overview of optical techniques for detecting wind profiles for wind energy applications is given in Emeis et al. (2007c). A special application result from a wind LIDAR specially developed for wind energy tasks is presented in Kindler et al. (2007). Here, a conically scanning LIDAR which uses beam focusing for range detection is used and tested.

Conical scanning with LIDARs for the determination of wind profiles can lead to problems in complex terrain, because this technique assumes horizontal homogeneity within the volume covered by the conical scan. However, in complex and mountainous terrain this assumption is often not valid. A first attempt for an estimation of the occurring error in such situations has been made from flow model simulations by Bingöl et al. (2009). They found errors in the order of 10%, but their model was the linearized model WAsP. So, this issue has to be investigated further with more sophisticated flow models.

Doppler LIDAR data at high resolution are also useful for characterizing turbulence properties of the atmosphere, either as mean statistics or as revealing turbulence events. Banta et al. (2006) defined vertical averaging bins for sequential vertical-slice scans and calculated mean and variance profiles of the streamwise wind component. These profile statistics were composited to relate mean and turbulence profile properties to those of the LLJ. Techniques for calculating other turbulence properties have been demonstrated, including second, third, and fourth moments of the vertical velocity (Lenschow et al. 2000) and velocity structure functions and coherence (Lothon et al. 2006).

Several methods have also been proposed for estimating the eddy dissipation rate from Doppler LIDAR data (e.g. Banakh and Smalikho 1997; Frehlich and Cornman 2002; Smalikho et al. 2005), which provide information on the fine-scale turbulence. Such fine-scale information is valuable for indicating overall turbulence levels or the presence of turbulent patches. Data from the high-resolution LIDAR HRDL have proven useful in probing the behaviour of many of these kinds of atmospheric flow features or events that produce turbulence. Breaking Kelvin–Helmholtz instabilities, for example, were studied by Newsom and Banta (2003), Blumen et al. (2001), and Poulos et al. (2002), and ducted gravity waves just above the LLJ maximum were studied by Fritts et al. (2003). Other types of discontinuity, including those with frontal or density current structure or larger-scale propagating gravity waves, were investigated by Darby et al. (2002a) and Sun et al. (2002, 2004).

A possible solution for the problem of large measurement volumes due to conical scanning with optical techniques may be the Dual-Doppler LIDAR technique. Here two identical LIDARs look into the same air volume from two different positions. Such an attempt using two 10.6 μm LIDARs is described in Collier et al. (2005). Newsom et al. (2005) used similar techniques to study boundary layer eddy structure near an urban centre. Both studies were directed at the improvement of urban dispersion modelling.

4.3.1.3 RADAR Windprofiler

UHF RADAR windprofilers operating at 915 or 1290 MHz can be used to monitor boundary layer and lower troposphere wind profiles with a vertical resolution of about 100 m. Boundary layer wind profile measurements together with extensive discussion of this monitoring technology are, e.g. reported in Ecklund et al. (1988) or Carter et al. (1995).

Figure 4.18 shows a typical time-height cross section of the horizontal wind from a UHF windprofiler (Caccia et al. 2004) for a layer between 400 and 3000 m above ground for a three-day period. The windprofiler was situated at the Mediterranean coast south of the French Alps. The figure displays an example for a mistral event characterized by north-westerly to northerly winds at this site. The mistral blows southward along the Rhone valley in Southern France through the gap between the Massif Central and the Alps (see also Section 4.5.4 below). Maximum winds of 30 m s^{-1} are observed on the first evening at 19.00 UTC. Wind profiles have been obtained every 15 min, but only one out of eight profiles is shown in the figure.

Observations of the vertical component of atmospheric turbulence, i.e. the variance of the vertical velocity component, have also been made with a newly developed low-power X-band boundary layer RADAR operating at 9.42 GHz (Al-Sakka et al. 2009).

4.3.2 Temperature

Air temperature is the most direct variable defining the static stability of the atmosphere and indicating the presence of inversions. Boundary layer temperature profiles can be obtained with active measurement methods such as RASS and Raman LIDAR. Additionally, passive methods using radiometer and interferometer measurements are available. The following subchapters will show examples.

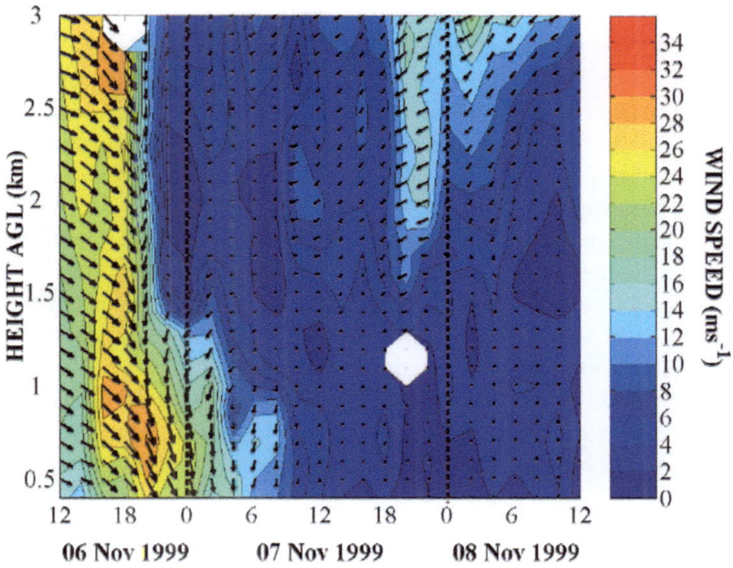

Fig. 4.18 Time-height cross section of horizontal wind from UHF windprofiler measurements. The *colours* code the wind speed, the *arrows* give the wind direction (from Caccia et al. 2004). Times are in UTC (roughly local time).*Vertical dotted lines* indicate midnight. *White areas* denote missing data

Besides RASS and Raman LIDAR, a third active technique is presently tested and may become available for the determination of tropospheric temperature profiles: high spectral resolution LIDAR (HSRL). This HSRL technique is resolving the temperature dependent line width of the Cabannes line, which is Doppler-broadened. This technique poses much higher demands in respect to the stability of the system as shown in Hair et al. (2001) than a Raman LIDAR (Radlach et al. 2008). Therefore, its operation is probably restricted to limited research projects.

4.3.2.1 RASS

Temperature measurements with the highest vertical resolution can be achieved with Doppler-RASS or SODAR-RASS measurements. This technique offers temperature profiles with about 20 m vertical resolution for several hundreds of metres above ground. Figure 4.5 shows an example of RASS measurements of potential temperature for a spring episode of three consecutive days. The diurnal variation from cool nights with stable stratification (potential temperature is increasing with height) and warm days with a well-mixed convective boundary layer (potential temperature is constant with height) can easily be identified. On the first day, the well-mixed period lasts from 2 p.m. to 7 p.m. CET (roughly local time minus one hour); on the other two days the well-mixed period is observable from noon to 7 pm. After 7 pm a new stable nocturnal surface layer starts to form, which persists throughout the night and the morning of the next day.

The white areas in Fig. 4.5, where temperature information is missing, provide an indirect evidence for the presence of a low-level jet in each of the covered four nights. A RASS is not able to observe any backscattered electromagnetic radiation if the sound pulse from the SODAR part of the instrument has been blown away from the focus of the electromagnetic antenna by too strong winds. With the instrument which was used to obtain the data in Fig. 4.5, this happened for horizontal wind speeds greater than about 12 to 15 m s^{-1}.

Figure 4.5 demonstrates the advantages of RASS as well as its disadvantages. Advantageous are the low first range gate just a few tens of metres above ground and the high vertical resolution. Disadvantageous are the limited height range, which usually is not sufficient to cover the entire vertical extend of the boundary layer and missing data for higher wind speeds. The determination of the temperature profile is not hampered by fog, clouds, or precipitation. Thus, for low to moderate wind speeds, a RASS offers a nearly one hundred percent data coverage. Therefore, the instrument can be used for monitoring purposes, especially for shallow and nocturnal boundary layers.

Due to the ability of a RASS to simultaneously monitor temperature and wind profiles, profiles of derived quantities such as the Brunt–Vaisala frequency and the Monin-Obukhov length can be analyzed as well. The Brunt–Vaisala frequency can be computed from the temperature profile over the whole measurement range and the Monin–Obukhov length from linear regressions of wind speed and potential temperature differences of the four lowest levels (Pérez et al. 2009).

4.3.2.2 Raman LIDAR

Rotational Raman LIDAR measurements deliver temperature profiles throughout the troposphere including the upper part of the ABL. The analysis is based on

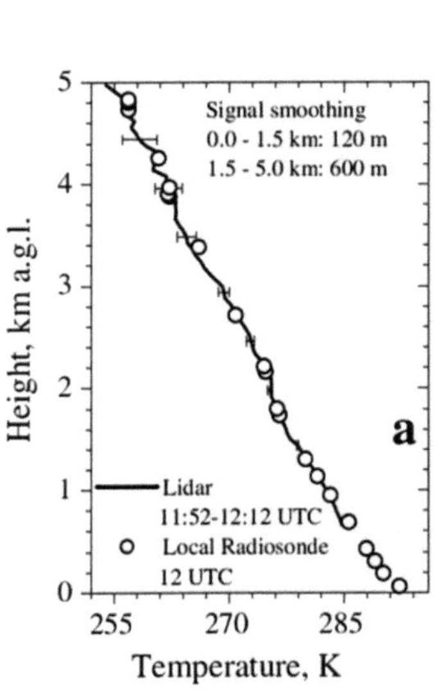

Fig. 4.19 Temperature profile from Raman LIDAR sounding (*full line*) compared to radiosonde data (*open circles*). From Radlach et al. (2008)

the temperature dependence of the intensity of the rotational Raman lines. Several ground-based Raman LIDAR systems exist today (Vaughan et al. 1993, Nedeljkovic et al. 1993, Behrendt and Reichardt 2000). Figure 4.19 shows an example obtained by the scanning rotational Raman LIDAR of the University of Hohenheim, a mobile system which is mounted on a truck (Radlach et al. 2008). The LIDAR data were acquired with a spatial resolution of 30 m and a temporal resolution of 60 s. The lowest range gate is about 600 m above ground. The scanning feature of the instrument allows for height cross sections of the temperature field.

A Raman LIDAR needs to be calibrated to obtain absolute temperature data. This is, e.g. done by comparison of the ratio of two rotational Raman signals with temperature profiles measured with radiosondes that are close in time and space to the LIDAR measurement. For 1 min integration time, an accuracy of the noontime temperature of better than 1 K can be achieved for ranges of less than 1 km (Radlach et al. 2008).

4.3.2.3 Passive Remote Sensing with Radiometers

This and the next subchapter demonstrate that passive remote-sensing techniques are suited for temperature profiling as well, if the terrain possesses sufficient horizontal homogeneity and some drawbacks in vertical resolution can be accepted. Temperature profiles (see Fig. 4.20 for an example) from scanning microwave radiometers at 5-mm wavelengths (60 GHz) gave excellent comparisons with in situ temperature measurements during winter conditions at the Boulder Atmospheric Observatory 300-m tower; several days of data were obtained during a snow storm, with no degradation of the quality of the data. A limited number of comparisons with a 915-MHz RASS above tower levels, showed the ability of the radiometer to derive temperature profiles to about 500 m height above ground with an accuracy of about 1°C (Westwater et al. 1999). The maximum range of these microwave radiometers is 600 m. The vertical resolution decreases drastically with height, being 7.5 m at 10 m, 65 m at 100 m, and 300 m at 400 m. Therefore, this passive radiometric technique is applicable only for the lower parts of the atmospheric boundary layer.

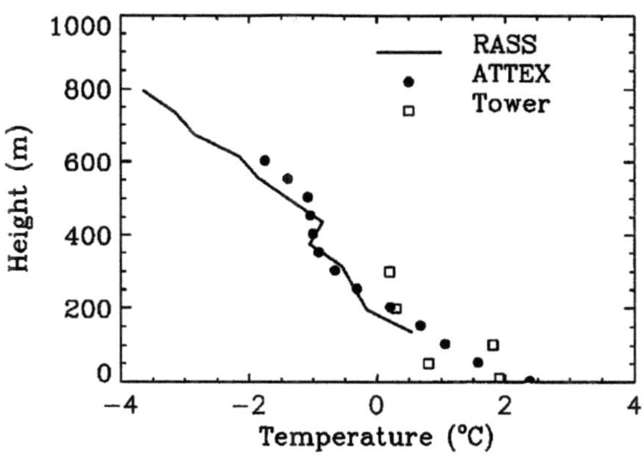

Fig. 4.20 Temperature profile from radiometer sonding (ATTEX, *bullets*) compared to tower (*open squares*) and RASS (*full line*) data. From Westwater et al. (1999)

The use of an enhanced radiometer, called HATPRO (Humidity And Temperature PROfiler), has been described by Pospichal and Crewell (2007). HATPRO records a set of 14 brightness temperatures every 2 s. From seven brightness temperatures from channels between 51.26 and 58 GHz, it obtains composite temperature profiles from boundary layer scans at six different elevations and zenith observations. These composite profiles will combine the advantages of both observation modes and give a consistent temperature profile from 50 m above ground to the upper troposphere. No significant differences are to be expected in the lowest kilometre, which is dominated by the boundary layer scans and above 2 km where all information stems from zenith observations. The vertical resolution of the temperature profile retrievals decreases from 10 m near the ground to about 300 m at 400 m height. Figure 4.21 gives an example from HATPRO temperature measurements in West Africa (Pospichal and Crewell 2007). In the lowest 500 m the retrieval accuracies for temperature from elevation-scanning microwave measurements with instruments such as HATPRO and spectral infrared measurements (such as AERI, see next subsection) are very similar (0.2–0.6 K). Above this level the accuracies of the AERI retrieval are significantly more accurate (<1 K RMSE below 4 km, Löhnert et al. 2009).

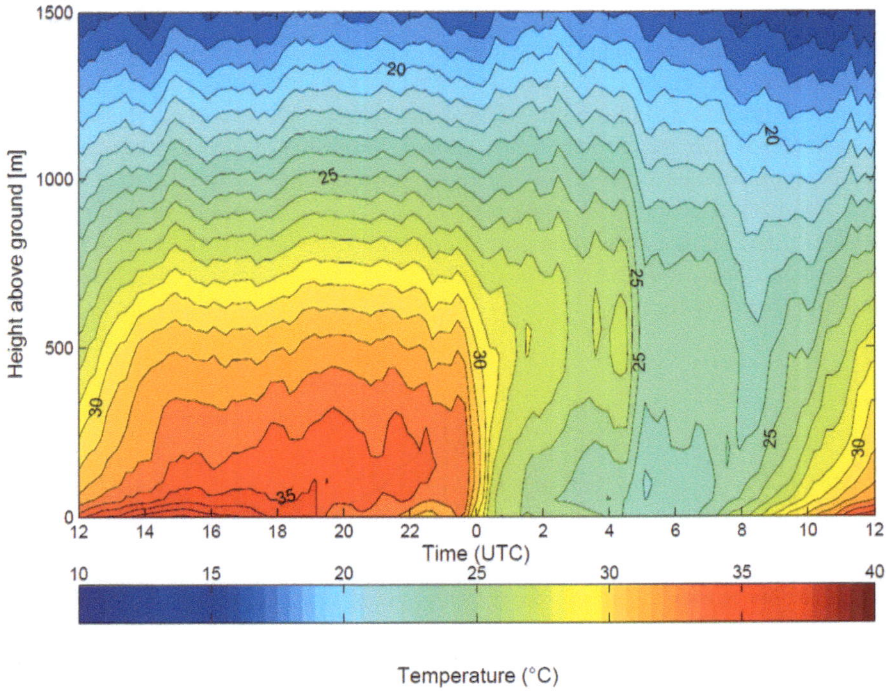

Fig. 4.21 24 hour time-height cross section of temperature from measurements with the radiometer HATPRO (from Pospichal and Crewell 2007)

4.3.2.4 Atmospheric Emitted Radiance Interferometer (AERI)

The very limited vertical resolution of passive remote sensing can be improved by obtaining spectra instead of measuring irradiances in certain spectral channels. An atmospheric emitted radiance interferometer (AERI) that delivers spectra can measure vertical temperature and humidity profiles with 10 min temporal and 100–200 m vertical resolution. Figure 4.22 shows an example presenting the passage of a cold front around noon and subsequent cooling especially in higher layers. The accuracy of the temperature retrievals has been analyzed to be better than 1 K (Feltz et al. 2003). Wind data from windprofiler measurements have been overlaid in Fig. 4.22 to show the wind direction shift towards North with the passage of the cold front. Above about 500 m above ground, instruments such as AERI are superior to microwave radiometers (Löhnert et al. 2009).

The German Weather Service operates an experimental version of a FTIR spectrometer at the Richard-Assmann observatory Lindenberg called EISAR (Emission Infrared Spectrometer for Atmospheric Research, Spänkuch et al. 1996, Engelbart 2005). This passive instrument records radiation from the frequency band between 3.3 and 20 μm.

4.3.3 Humidity, Water, and Ice

This section deals with active measurements of gaseous humidity as well as of precipitation elements. Atmospheric water vapour is the most abundant trace gas in the Earth's atmosphere. Due to the possible latent heat release during phase changes of water vapour into liquid and solid states, atmospheric humidity has influence on atmospheric thermal stability. Due to this and its influence on the atmospheric energy budgets, humidity is treated here separately from other trace gases. Gaseous humidity can actively be detected by DIAL and Raman LIDAR observations and passively by radiometer and interferometer measurements. All other trace gases are addressed in the subsequent section.

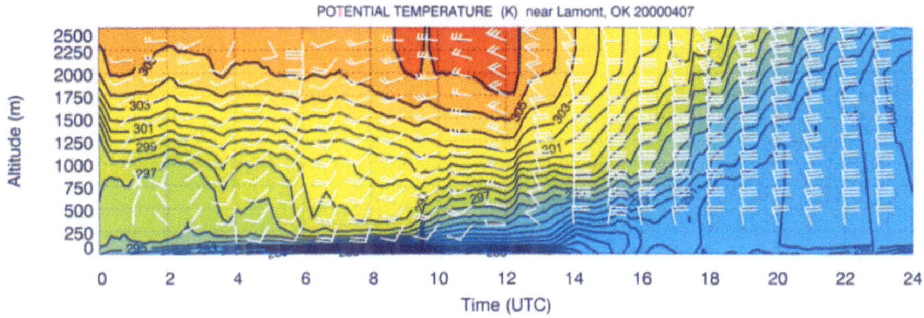

Fig. 4.22 Time-height cross section of potential temperature from AERI measurements (*colours* and *lines* with labels). *Horizontal wind arrows* are from additional 404 MHz wind profiler soundings (from Feltz et al. 2003)

For a long time, rain, snow, and hail have already been monitored by weather RADARs. RADARs give boundary layer profiles only in the near range if ground clutter is not too high. The vertical resolution is quite coarse due to the large opening angle of the RADAR beam. Here, we will address with respect to precipitation only measurements with vertically looking micro rain RADARs.

4.3.3.1 DIAL Measurements of Humidity

Atmospheric humidity profiles can be obtained from DIAL measurements. A suitable DIAL could, e.g. consist of a alexandrite ring laser that is injection-seeded by a continuous wave Ti:sapphire ring laser. The laser beam is expanded 15-fold and directed vertically into the atmosphere. The direction of the laser beam was aligned to achieve full overlap with the field of view of the detector system at about 400 m. (Wulfmeyer 1999). This instrument has a lowest range gate of 200 m, a temporal resolution of 10 s, and a vertical resolution of 30 m. The accuracy for absolute humidity in the entire troposphere up to 7000 m is better than 5%. Figure 4.23 displays an example for such a DIAL measurement of absolute humidity from Wulfmeyer (1999). The top of the more humid boundary layer is between 400 and 800 m above ground in the shown profiles. Above the boundary layer the atmosphere is drier. The zone between 400 and 800 m is interpreted as an entrainment zone from additional turbulence measurements (Wulfmeyer 1999).

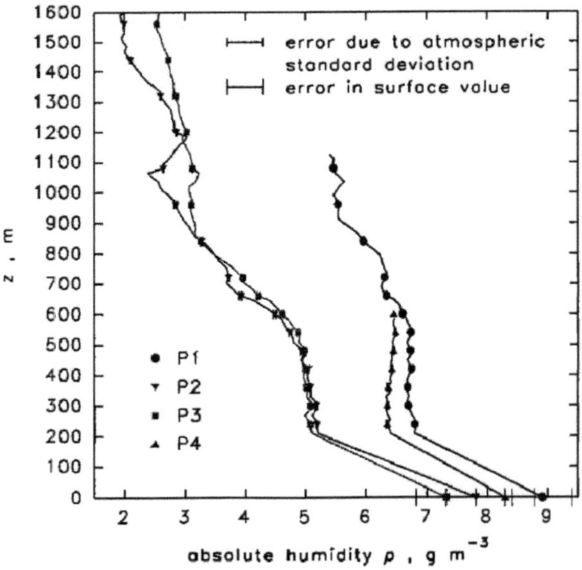

Fig. 4.23 Absolute humidity profile from DIAL measurements (data points above 200 m) for four different time periods, each roughly one hour in duration. Linear interpolation has been performed between the lowest DIAL data and surface data. From Wulfmeyer (1999)

4.3.3.2 Raman LIDAR Measurements of Humidity

Alternatively, specific humidity or mixing ratio can be determined from Raman LIDAR measurements by taking the ratio of the signals from the first Stokes vibrational Raman shifts for water vapour, S_{H2O} and nitrogen, S_{N2}. The measurement is made with laser transmission at visible (532 nm) and ultraviolet (266 nm) wavelengths. The visible measurement (660/607 nm) is available at night and the ultraviolet measurement (294/284 nm) is available day and night. The ultraviolet measurement is limited to the first 3 km of the atmosphere because of signal loss due to the large scattering cross section. The specific humidity $q(z)$ is determined from the relationship,

$$q(z) = K\frac{S_{H2O}(z)}{S_{N2}}, \tag{4.14}$$

where the calibration factor, K, is verified by comparison with radiosonde measurements when the instrument is moved to a new site (Philbrick and Mulick 2001). Figure 4.24 shows an example of a humidity profile obtained with LAPS (Pennsylvania State University Laser Atmospheric Profile Sensor, Philbrick 2002). The humidity profile data from the Raman LIDAR coincide well within 10% with

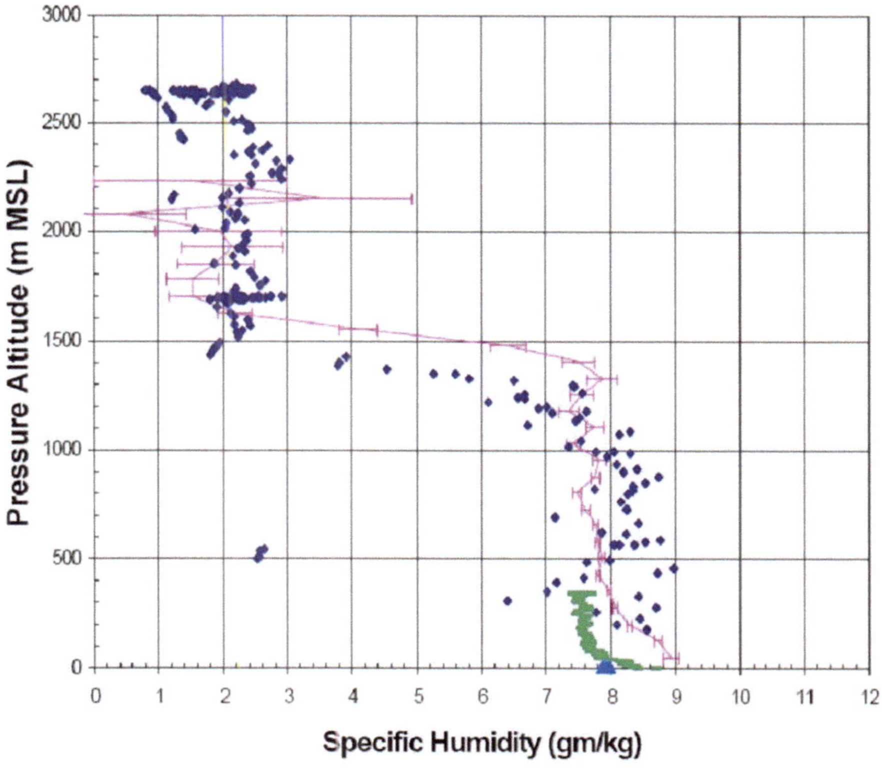

Fig. 4.24 Specific humidity profile from Raman LIDAR measurements (*diamonds*) with LAPS (Lidar Atmospheric Profile Sensor) compared to thethered balloon measurements (*bold line* in the lower 400 m) and aircraft measurements (*thin full line*). From Philbrick (2002)

tethered balloon and aircraft data. Inspection of the profile suggests that the top of the boundary layer is around 1200–1400 m above ground where the specific humidity drops drastically with height.

For many purposes the knowledge of relative humidity is more important than specific humidity or mixing ratio. Relative humidity controls, e.g. cloud formation and aerosol optical properties and thus visibility. Mattis et al. (2002) have, therefore, tried to directly determine relative humidity profiles of the air from Raman LIDAR measurements. For this purpose they combined in one instrument the Raman-LIDAR techniques that are used for the profiling of water vapour and temperature. Close agreement of the remotely measured profiles of relative humidity and those measured with radiosonde demonstrates the potential of this approach. A vertical resolution of 100 m in the boundary layer and a relative error of less than 10% seem achievable from this study.

4.3.3.3 Radiometer Measurements

The radiometer HATPRO (Pospichal and Crewell 2007), which is able to retrieve temperature profiles (see Section 4.3.2.3), can be used to obtain boundary layer humidity profiles, too, from an analysis of the signals from the seven channels between 22.24 and 31.4 GHz. In contrast to the temperature soundings, no boundary layer scans are performed for obtaining humidity profiles. Therefore, the vertical resolution is several hundreds of meters only. Profiles shown in Rose et al. (2005) demonstrate that the error in absolute humidity measurements is about 10%. Like the Raman LIDAR described just before, HATPRO is able to simultaneously observe temperature and humidity profiles as well. This again allows for the determination of relative humidity profiles in the boundary layer and the whole troposphere.

4.3.3.4 Atmospheric Emitted Radiance Interferometer (AERI) and FTIR

An atmospheric emitted radiance interferometer (AERI) can measure vertical temperature and humidity profiles with 10 min temporal and 100–200 m vertical resolution. Figure 4.25 shows the passage of a dry line as an example. The accuracy of such humidity retrievals is approximately 5% of the absolute value when compared with well-calibrated radiosonde data. The parallel retrievals of temperature and moisture profiles with this instrument allow for the assessment of the onset of deep convection and the formation of thunderstorms (Feltz et al. 2003). AERI retrievals show high potential, especially for retrieving humidity in the boundary layer, where accuracies are on the order of 0.25–0.5 g m^{-3} for a central European climate (Löhnert et al. 2009).

Due to a large vertical gradient and strong variability of water vapour, algorithms that are effectively applied for ground-based remote sensing of many different atmospheric trace gases (see below) can be insufficient for the retrieval of tropospheric water vapour profiles. A review of the most important features of the retrieval and of the radiative transfer modelling required for accurate monitoring of tropospheric

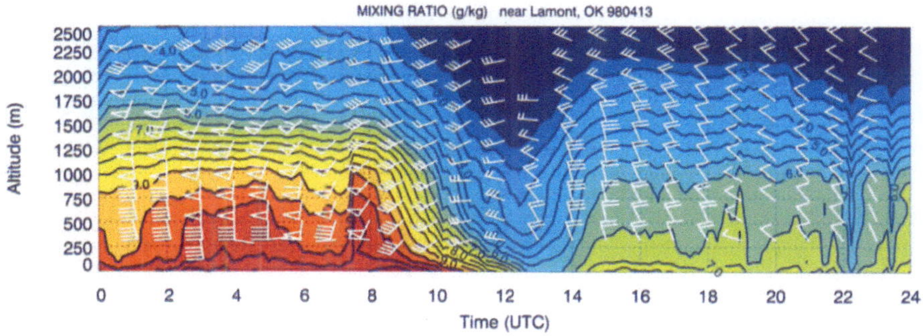

Fig. 4.25 Time-height cross section of mixing ratio from AERI measurements (*colours* and *labelled lines*). *Horizontal wind arrows* are from additional 404 MHz wind profiler soundings (from Feltz et al. 2003)

water vapour profiles by ground-based FTIR (Fourier Transform Infrared) experiments is given in Schneider and Hase (2009). The methods discussed in this review are: a fit of a variety of different water vapour lines with different strength, a logarithmic scale inversion, a speed-dependent Voigt line shape model, and a joint temperature profile retrieval.

4.3.3.5 Micro Rain RADAR Measurements

So far, measurements of gaseous humidity have been presented. Figure 4.26 now gives an example for micro rain RADAR measurements. Figure 4.26a shows a waterfall presentation of the spectral RADAR reflectivity versus Doppler frequency

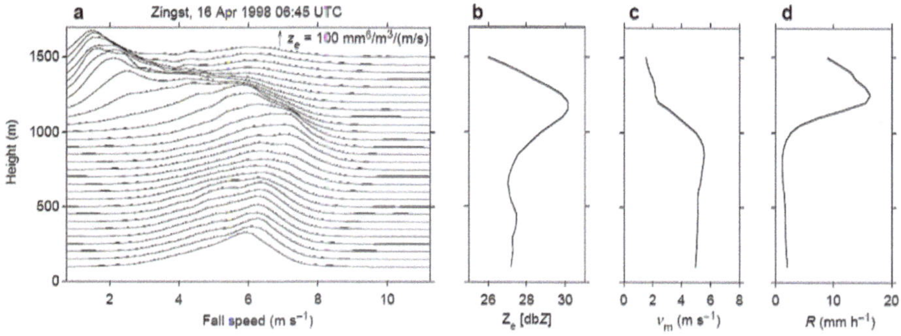

Fig. 4.26 Micro rain radar measurement example. **a**: Waterfall presentation of raw Doppler spectra of radar reflectivity averaged over 1 min from 100 to 1500 m. The height resolution is 50 m. The z_e-calibration is given by the *arrow* in top of the figure. — **b**: Profiles of radar reflectivity Z_e. — **c**: Mean fall velocity v_m. — **d**: Rain rate R. — The melting level at 1100 m appears as a peak of Z_e (bright band), as a step of v_m and as a peak of the apparent rain rate R, respectively. From Peters et al. (2002)

for all range gates in one selected minute from measurements in April 1998. The spectral peak in the lower range gates is at about 6 m s^{-1}. At heights above 1100 m (the melting level at this time), the spectral peak is shifted to 2 m s^{-1} and becomes fairly narrow which is typical for snowfall. The corresponding profiles of RADAR reflectivity, mean fall velocity, and rain rate are shown in Fig. 4.26b–d. The melting level appears in these profiles as enhanced reflectivity, as step in fall velocity, and as apparently enhanced rain rate, respectively. The last feature is an artefact, as the retrieval is only applicable for the liquid phase. Nevertheless, it can be used as sensitive indicator for the melting level, even when it is not detectable in the reflectivity profile. These profiles, showing the melting level, may be used to identify the state of precipitation, rain, or snow. This information could be helpful also for the interpretation of weather RADAR data (Peters et al. 2002).

By assuming a relationship between drop size and the vertical velocity, the drop size distribution can be derived. A detailed error analysis has been performed by Peters et al. (2005) on a theoretical basis and by comparison with surface disdrometers. It was found that the deviations of the micro rain RADAR drop size distributions from the in situ drop size distributions are in the same range as the mutual deviations of the in situ drop size distributions (Crewell et al. 2008).

4.3.4 Trace Substances

Remote sensing of other trace substances than water vapour can be done along vertically and horizontally oriented paths in the atmospheric boundary layer. Only optical techniques are available for these gases and aerosols. Usually, LIDARs such as DIAL, Raman LIDARs, and ceilometers are operated in a vertical profiling mode while FTIR and DOAS are employed for horizontally path-averaged trace gas measurements.

Remote sensing of trace substances may be based on three methods, either the detection of radiation absorption, emitted radiation, or by scattered radiation. The wavelength-dependent absorption of radiation by trace gases may be utilized to measure the concentration of trace gases in the atmosphere. Depending on the physical properties of the gas molecules (i.e. the typical energy amounts these molecules can absorb by electronic transitions, excitations, and increased vibrational and/or rotational energy), different spectral ranges must be analyzed to find these gases. For the detection of trace gases, analyses are made either at one single absorption line or over a wider range of spectra. For investigations concentrating on one absorption line, interference from other trace gases can become important and should be known. On the other hand, the spectral signature or fingerprint of a trace gas in a wider spectral range is rather unique. For spectral methods splitting of incoming radiation can be achieved either by a grating or prism spectrometer, or by in an interferometer.

In the UV range, we find, e.g. the absorption bands of ozone molecules (Hartley bands between 200 and 300 nm, a Huggins band at 340 nm, see Malicet et al. 1995).

Oxygen (O_2) and nitrogen (N_2) molecules also have absorption bands in the UV range. O_2, H_2O, NO_2, ClO_2, and BrO have absorption bands in the visible range. All these absorption bands originate from electronic transitions.

In the infrared range, combined rotational and vibrational energy changes lead to absorption bands. The absorption frequency depends on the vibration frequency of the molecules. All atmospheric gases except diatomic gases have absorption bands in the infrared range and thus also contribute to the greenhouse effect, which keeps the Earth warm.

In the far infrared (> 40 μm), molecules with a permanent dipole moment have rotational absorption bands. These gases include H_2O, O_3, OH, CO, HCl, HF, HBr, N_2O, HCN, HOCl HO_2, HNO_3, and ClO.

In the microwave range (>1 mm), rotational bands of H_2O, O_3, H_2O_2, ClO, and HNO_3 can be utilized for the detection of these trace gases. The following subsections give a few examples.

4.3.4.1 Vertical Profiling

Ozone profiles in the lower troposphere can be measured with a Raman LIDAR using a DIAL analysis of the ratio of the vibrational Raman signals for nitrogen (284 nm) and oxygen (278 nm), which are on the steep side of the Hartley band of ozone (Philbrick and Mulik 2001). Figure 4.27 shows an example which reveals a threefold vertical structure of the ozone concentration in that night. There is a shallow surface layer with very low concentrations due to ongoing titration of ozone with freshly emitted nitric oxides from car traffic. A second layer with a uniform concentration level extends from 200 to 1700 m. This is most probably a residual layer which was left from the CBL of the preceding day. Above this residual layer, even higher ozone concentrations can be found in the free troposphere. In the free troposphere, ozone patches can exist for relatively long period of time and are frequently observed to drift in the background wind, as observed in this case (Philbrick and Mulik 2001).

The left frame in Fig. 4.3 displays an example for a time-height cross section of the optical backscatter intensity sampled by a ceilometer with one optical axis (the emitted light beam is sent through a hole in the receiving mirror). The backscatter is similarly caused by aerosol particles and water droplets floating in the air. The aerosol information from such a backscatter LIDAR is qualitatively only, because the particle size distribution is not known. Furthermore, the backscatter intensity may be modified by the atmospheric humidity as well, if the particles are hygroscopic. The figure shows high backscatter intensities due to mist or weak fog in the surface layer in the night and early morning hours (yellow to red colours in the lower left of Fig. 4.3). Above this foggy layer a second layer with enhanced aerosol concentrations is visible. This layer is topped at 1200– 1400 m. At about 8 a.m., the formation of a convective boundary layer (CBL) starts which rapidly gains height. In the afternoon the maximum vertical extension of this CBL is reached with about 1400 m. The air within the CBL seems to be nearly perfectly mixed, because the backscatter intensity is very even throughout the whole CBL. The shown backscatter

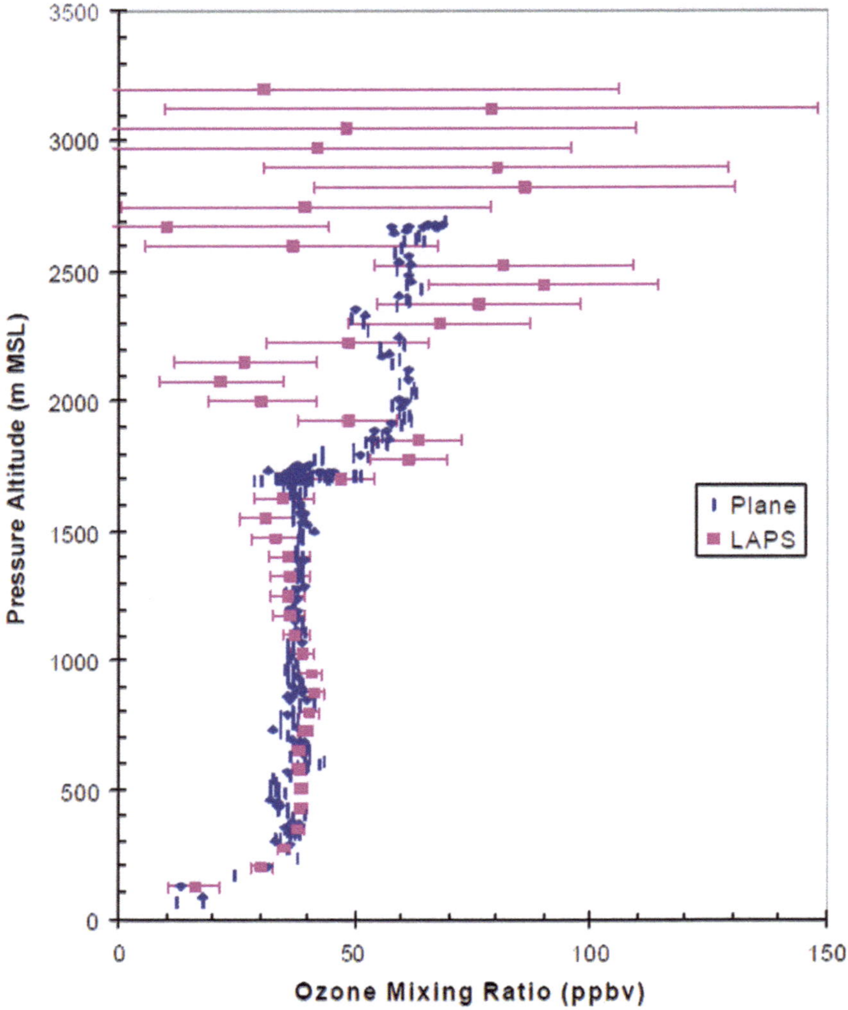

Fig. 4.27 Nocturnal ozone mixing ratio profile over Philadelphia in the early hours of August 20, 1998, from Raman LIDAR measurements (*squares* with error bars) with LAPS (Lidar Atmospheric Profile Sensor) compared to aircraft measurements (small symbols). From Philbrick and Mulik (2001)

pattern is typical for a clear spring or early summer day with a radiation-dominated diurnal variation of the boundary layer structure.

4.3.4.2 Horizontally Path-Averaging

Path-averaging measurements of trace gas and pollutant concentrations in the infrared range can be made with FTIR and in the UV and visible range with

DOAS. These measurement techniques allow for temporally continuous observations. Path-averaged data are obtained for monitoring purposes as well as for evaluation of numerical model results. For the latter purpose path-averaged data are much better suited than punctual in-situ measurements, because they are averaged over length scales that are comparable to the grid distances of the numerical models.

With FTIR CO, CO_2, CH_4, N_2O, NO, NO_2, SO_2, NH_3, SF_6, HCHO, BTX (benzine, toluene, xylene), as well as *n*-pentane, ethane, ethene, ethine, propane, propene, (iso-)butane, iso-butene, butadiene, and hexane can be detected. To interpret infrared spectra of atmospheric measurements, a multi-component air pollution software (MAPS) was developed for retrieval of gas concentrations from radiation emission as well as absorption measurements (Schäfer et al. 1995). In absorption mode, a radiation source is needed (either the sun or an artificial heat source). The upper left frame in Fig. 4.8 shows a methane concentration time series obtained from FTIR measurements.

FTIR in emission mode does not need a radiation source. Thus imaging by scanning is possible. The scanning infrared gas imaging system (SIGIS-HR), which is based on the FTIR technology, together with the quantitative gas analysis software MAPS offers the investigation of the spatial distribution of the temperature and gas concentrations (e.g. CO, NO) within hot plumes of aircraft engines at airports. The system integrates an infrared camera as well for the localization of the hot source that additionally suggests the best measurement position of the SIGIS-HR (Flores et al. 2007).

With DOAS, BTX, and other reactive gases such as O_3, NO_2, NO_3, (OH), HCHO, HONO, and SO_2 can be detected. A comparison of different DOAS instruments is presented in Camy-Peyret (1996). Recent advances in tropospheric chemistry observation by spatially resolving spectroscopic techniques like active and passive DOAS tomographic measurements of two-dimensional trace gas distributions, as well as ground-based and airborne Imaging DOAS (I-DOAS) observation of 2D and 3D trace gas patterns, are described by Platt et al. (2009). A particularly promising approach is the combination of tomographic techniques with imaging – DOAS on airborne platforms, which can provide three-dimensional trace gas distributions.

An example for the determination of vertical gradients of NO_3, N_2O_5, O_3, and NO_x concentrations in the nocturnal boundary layer with DOAS measurements along slanted paths is given in Stutz et al. (2004). Two DOAS instruments were mounted in 2 m height on an airport and five different retroreflector arrays were deployed in some distance at different heights, with the goal of measuring the vertical distribution of the trace gases. Three arrays were mounted on a radio tower at 6.1 km distance from the DOAS telescopes at 70 m, 99 m, and 115 m height above the ground. Another retroreflector array was mounted on a water tower at 1.9 km distance and an altitude of 44 m. The last and shortest light path ran 2 m above the meadow inside the airport. Results from this experiment are vertical trace gas profiles over the first 115 m above ground with roughly 20 m vertical resolution. These profiles are horizontally averaged over the DOAS path lengths.

DOAS can also be used for emission studies. For example, a DOAS consisting of a emitter/receiver unit and three retroreflectors was used for a motorway pollutant emission study. One path was oriented across the motorway (120 m path length) in about 10 m altitude above street level, another path was set up in parallel to the motorway, and the third path was operated perpendicularly away from the motorway. Comparison of the concentrations found along these three paths allows for an estimation of the pollution emission from the motorway (Schäfer et al. 2006b).

4.4 Vertical Flux Profiles

Fluxes of momentum, energy, and substances are important links between the different compartments of the Earth system. Fluxes from the atmosphere to the Earth's surface and vice versa must pass through the ABL and have considerable influences on the state of the ABL. Fluxes usually consist of a mean and a turbulent part. The mean part happens with mean atmospheric motions while the turbulent fluxes are due to the covariance between turbulent fluctuations of the wind and the respected transported quantity without any mean motion. We will focus here on vertical fluxes through the ABL. Due to the proximity of the Earth's surface mean vertical motions in the ABL are usually small, so that turbulent fluxes are the dominant fluxes in the ABL. The direct measurement of turbulent fluxes requires the simultaneous observation of vertical wind fluctuations and fluctuations of the respective transported quantity with high temporal resolution of at least 10 Hz. Such observations are already difficult with in situ instrumentation, the more this is problematic with remote-sensing instrumentation. Nevertheless, quite considerable efforts have been made in the past decades to detect vertical profiles of turbulent fluxes in the ABL (Table 4.3).

An overview of the measurement of turbulent momentum and heat fluxes with wind profiler RADAR, SODAR, and RASS has been given by Engelbart et al. (2007). Tables 4.4 and 4.5 have been updated from this review. Although they see impressive advances in the development of methods for remote determination of turbulent fluxes in the ABL, no presently known method can be used already for routine measurements. Therefore, the following subsections on remote flux measurements can only serve as references to the current efforts to work on this issue.

Table 4.3 Overview on methods to derive vertical profiles of turbulent fluxes using ground-based remote sensing (see rightmost column for section number)

Variable	Detection method	Section
Momentum	K- and X-band RADAR, Doppler wind LIDAR	4.4.1
Heat	RASS, scintillometry	4.4.2
Humidity	DIAL plus Doppler wind LIDAR or wind profiler	4.4.3

Table 4.4 Direct measurements of the momentum flux using profiling techniques (updated from Engelbart et al. 2007)

	Device	Reference	Res./range	Examination	Restriction
RADAR	Scanning and dual, 10 GHz	Rabin et al. (1982) Kropfli (1986)	~100/200–2000 m	Isolated comparison with mast and aircraft	Shortage of scatterers
	Scanning, 10 GHZ + chaff	Schneider and Lilly (1999)	~100/200–3500 m	Comparison with mast and aircraft	Chaff air contamination
	Five-beam, 1.238 GHz	Puygrenier et al. (2004)	75–2000 m	Comparison with aircraft	Ground clutter
LIDAR	Scanning	Eberhard et al. (1989) Gal-Chen et al. (1992)	~100/400–1500 m	Tested	Integration time of ~1.5 hours
	Scanning and dual	Levinson (1998) Middleton and Davies (2004)	~100/400–1500 m	Used in urban turbulence studies	Integration time of ~1.5 hours
	Scanning	Pichugina et al. (2005b) Banta et al. (2006) Smalikho et al. (2005)	~30/50–300 m	Comparison with mast, used in low-level jet studies	
SODAR	Phased-array system	Helmis et al. (2004)	40/50–400 m	Tested	Low sampling rate
	Vertically pointing	Kouznetsov et al. (2004) Kouznetsov et al. (2007)	20–400 m	Experimental	Additional assumptions needed
RADAR-RASS	915 MHz	Angevine et al (1993a,b)	~100/150–900 m	Comparison with aircraft	
	1235 MHz	Peters and Kirtzel (1994) Lippmann et al. (1996) Hirsch and Peters (1998)	~40/120–800 m	Tested	Special measurement mode

Table 4.5 Direct measurements of the heat flux using profiling techniques (updated from Engelbart et al. 2007)

	Device	Reference	Res./range	Examination	Restriction
RADAR-RASS	482 MHz	Vogt and Engelbart (2002)	120/460–900 m	Comparison with other techniques	
	915 MHz	Potvin et al. (1998) Potvin and Rogers (1999)	~100–1000 m	Comparison with aircraft	
	1235 MHz	Peters and Kirtzel (1994) Lippmann et al. (1996) Hirsch and Peters (1998)	~40/120–800 m	Tested	
	1290 MHz	Engelbart and Klein Baltink (1997) Engelbart (1998) Engelbart et al. (2002) Engelbart and Bange (2002)	100/200–900 m	Comparison with tower, helicopter	
SODAR-RASS	1 GHz	Peters et al. (1985) Engelbart et al. (2002) Engelbart and Bange (2002) Vogt and Engelbart (2002)	~20/40–500 m	Tested, comparison with mast	Low sampling rates
LIDAR	Doppler-LIDAR	Davis et al. (2009)	100–800 m	Experimental	Assumptions necessary

4.4.1 Momentum Flux

K- and X-band RADARs have been used first to obtain turbulent vertical fluxes of horizontal momentum (Kropfli 1986). For instance, he used insects and seeds in a summertime convective boundary layer as tracers of air motion. He similarly used these RADARs to observe the momentum flux in a non-precipitating cloud.

Vertical profiles of turbulence and momentum fluxes have been derived from ground-based remote sensing with high-resolution pulsed Doppler LIDAR instruments for the first time by Eberhard et al. (1989). Ubiquitous small aerosol particles serve as tracers for the air motion. The LIDAR performs conical scans to detect all three components of the wind vector. The valuation is then done in three steps. First, the mean wind components are determined. Then, the variance of the measured radial velocities is determined as a function of azimuth. Finally, using partial Fourier decomposition, turbulence parameters are extracted from the azimuth-dependent variance under the assumption of horizontally homogeneous statistics of the turbulence (Eberhard et al. 1989). Figure 4.28 shows an example with an exponentially decreasing momentum flux with height in a convective boundary layer. The flux nearly vanishes at the top of the CBL. The s-shape near the top of the CBL is not reliable (see error bar). The vertical resolution of the shown flux profile is about 150 m.

A different method tries to derive momentum fluxes from SODAR measurements. This method is based on the assumption that there is in the lower part of the neutrally stratified boundary layer a simple relationship between the momentum flux $\overline{u'w'}$ and the variance of vertical wind component σ_w^2 (Kouznetsov et al., 2004):

$$\overline{u'w'} = -0.77\sigma_w^2. \tag{4.15}$$

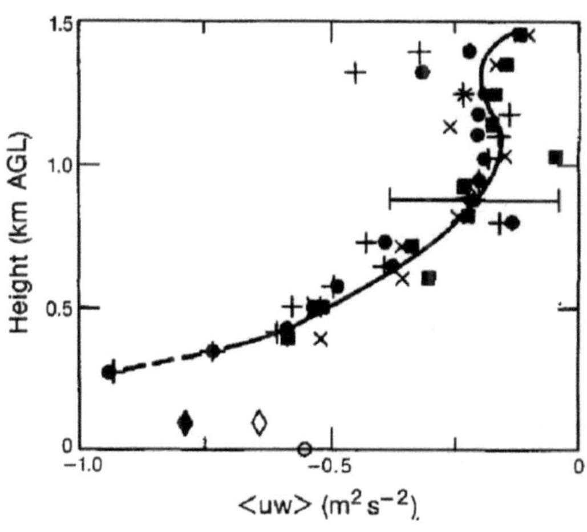

Fig. 4.28 One-hour mean momentum flux profile compared to aircraft measurements (diamonds). From Eberhard et al. (1989)

This relationship (4.15) was derived from the data taken at altitudes 50 to 100 m above ground level. Numerous data taken in the surface layer give slightly smaller value of the coefficient: 0.62–0.7 (see e.g. Stull 1988). Using this relationship one can estimate the vertical structure of momentum flux in neutrally stratified ABL by means of a SODAR (or any other remote-sensing tool) that is able to measure σ_w^2 accurately enough (Kouznetsov et al. 2007). Figure 4.29 shows empirical data which give the dependence of the proportionality factor in (4.15) as function of the flux Richardson number, R_f. Mandatory input for the determination of the flux Richardson number, i.e. the heat flux, cannot be observed from SODAR measurements, but must be specified from separate surface heat flux measurements and the assumption that the heat flux decreases linearly to zero at the top of the mixing layer (Kouznetsov et al. 2007).

Basically the same idea, on which (4.15) is founded, can be employed to estimate vertical profiles of the turbulent viscosity, ν_t or eddy viscosity of the air (Emeis 2004; Kouznetsov et al. 2007). This turbulent viscosity is an important parameter in turbulence parameterization schemes for meso-scale numerical flow models. Frequently, this turbulent viscosity is termed turbulent vertical exchange coefficient, $K_{m,v}$. It is defined as

$$K_{m,v} = \nu_t = -\frac{\overline{u'w'}}{\frac{\partial u}{\partial z}}, \tag{4.16}$$

As the vertical turbulent momentum flow is much larger than the mean vertical momentum flow, this is the most important component that has to be parameterized in the numerical simulation of turbulent boundary layer flow. Applying (4.16) to a homogeneous Prandtl-layer (i.e. putting $u*^2$ for $-\overline{u'w'}$ and $u*/\kappa z$ for $\frac{\partial u}{\partial z}$, with the von Kármán constant $\kappa = 0.4$) leads to the well-known expression for the turbulent exchange coefficient $K_{m,v}(=\nu_t)$ that increases linearly with height if the wind speeds increases logarithmically with height:

$$\nu_t = u * \kappa z, \tag{4.17}$$

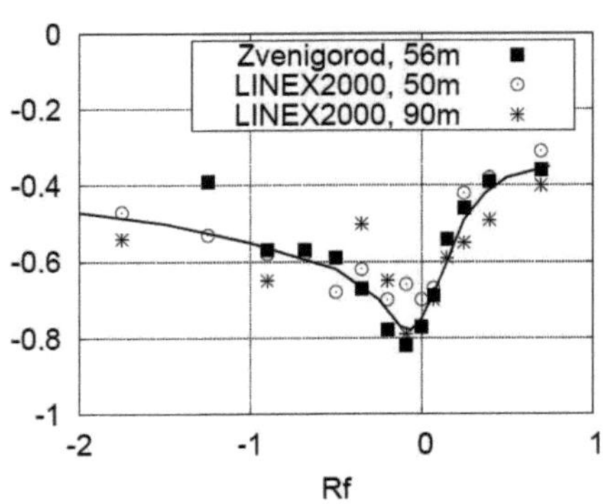

Fig. 4.29 Factor of proportionality between the turbulent vertical momentum flux and the square of the standard deviation of the vertical wind component as function of atmospheric stability expressed in terms of the flux Richardson number, R_f. From Kouznetsov et al. (2007)

Using (4.16), we can determine the turbulent viscosity in a real flow if we measure the turbulent momentum flux $\overline{u'w'}$ and the shear $\frac{\partial \overline{u}}{\partial z}$ as it has been done by Campistron et al. (1991). While the shear as a mean quantity is relatively easily available from SODAR measurements, it is very difficult to get the turbulent momentum flux from remote-sensing data (see above).

Looking for a surrogate for the turbulent momentum flux, Emeis (2004) proposes to use the findings from Stull (1988) or, e.g. from data presented in Emeis et al. (1995) that ${\sigma_w}^2$ and $\overline{u'w'}$ are proportional in the surface layer. In the case of near neutral and stable thermal stratification of the air over homogeneous terrain, this gives

$$\sigma_w^2 = 1.6\overline{u'w'}, \tag{4.18}$$

Inserting this proportionality into (4.16) we get

$$\nu_t = \frac{\sigma_w^2}{1.6\frac{\partial \overline{u}}{\partial z}}, \tag{4.19}$$

(4.19) is valid only for a near neutral or stable surface layer. Above this layer the situation is more complicated. For a near neutral mixed layer where mechanical production of turbulence dominates, both σ_w and the turbulent momentum flux decrease with height. Using again information from Stull (1988, his equations 9.5.2b, 9.5.3f, and 9.5.3j), we get with the mixing layer height z_i (this is identical to MLH above)

$$\sigma_w^2 = 1.6\left(1 - \frac{z}{z_i}\right)^{-0.5} \overline{u'w'}, \tag{4.20}$$

which, if the factor 1.6 from Stull's (1988) equation 9.5.3j is used in his equation 9.5.3f converges to (4.18) for z against zero. For $z = 0.5\ z_i$ the factor between ${\sigma_w}^2$ and $\overline{u'w'}$ is 2.26, for $z = 0.9\ z_i$ it is 5.06. By inserting (4.20) into (4.16), this leads to an approximation for ν_t

$$\nu_t = \frac{\sigma_w^2}{1.6(1 - \frac{z}{z_i})^{-0.5}\frac{\partial \overline{u}}{\partial z}}, \tag{4.21}$$

(4.21) should not be used very close to the top of the mixed layer. For a stable mixed layer, equations 9.4.2c and 9.4.3f in Stull (1988) suggest a more or less height independent, constant ratio between ${\sigma_w}^2$ and $\overline{u'w'}$ of the order 2.5. Thus, a rough compromise between this finding, eq. (4.19), and eq. (4.21) must be used

$$\nu_t = \frac{\sigma_w^2}{a\frac{\partial \overline{u}}{\partial z}}, \tag{4.22}$$

with a height-dependent a. Emeis (2004) puts $a = 1.6$ in a lower layer that includes the surface layer (he chooses 55–200 m for this layer), $a = 2.0$ for a middle layer (200–600 m), and a = 2.5 for an upper layer (600–1000 m). The approximation (4.22) to turbulent viscosity (4.16) has been made because (4.22) can be computed from SODAR data without the need of additional information. This offers the opportunity to determine the diurnal course and the vertical structure of turbulent viscosity and to estimate the range of values this quantity can take from longer time series of ground-based measurements with a SODAR. Because a quadratic expression (σ_w^2) instead of $\overline{u'w'}$ is used in (4.22), the information about the sign of the turbulent momentum flux is lost in (4.22). In order to assure the positiveness of the turbulent viscosity, only the absolute value of the shear must be used in (4.22).

Figure 4.30 shows a time series of the turbulent viscosity for two different heights above ground over 12 days together with $\overline{\frac{\partial u}{\partial z}}$ and σ_{w}, which have been used to derive this variable. The most prominent feature in Fig. 4.30 is the daily cycle of the magnitude of the turbulent viscosity with a maximum around noon and a minimum around midnight. This daily cycle is not only due to such a cycle in σ_{w} but especially in

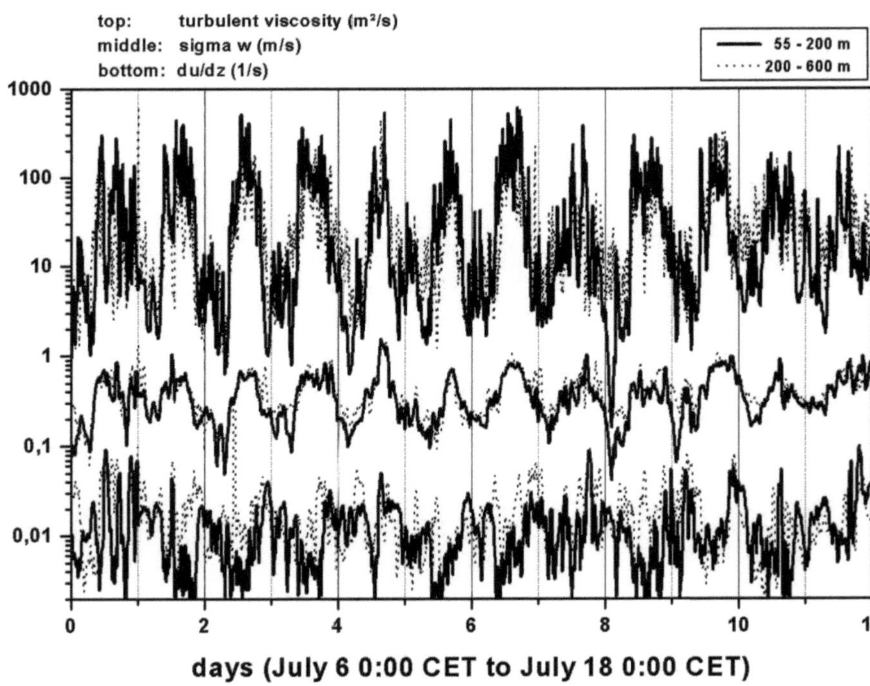

Fig. 4.30 Time series of turbulent viscosity in m^2 s^{-1} (top), standard deviation of the vertical wind component in m s^{-1} (middle) and vertical shear of the horizontal wind in 1 s^{-1} (below) for two height ranges (55–200 m above ground: *full lines*), 200–600 m above ground: *dashed lines*) from SODAR measurements using eq. (4.22).

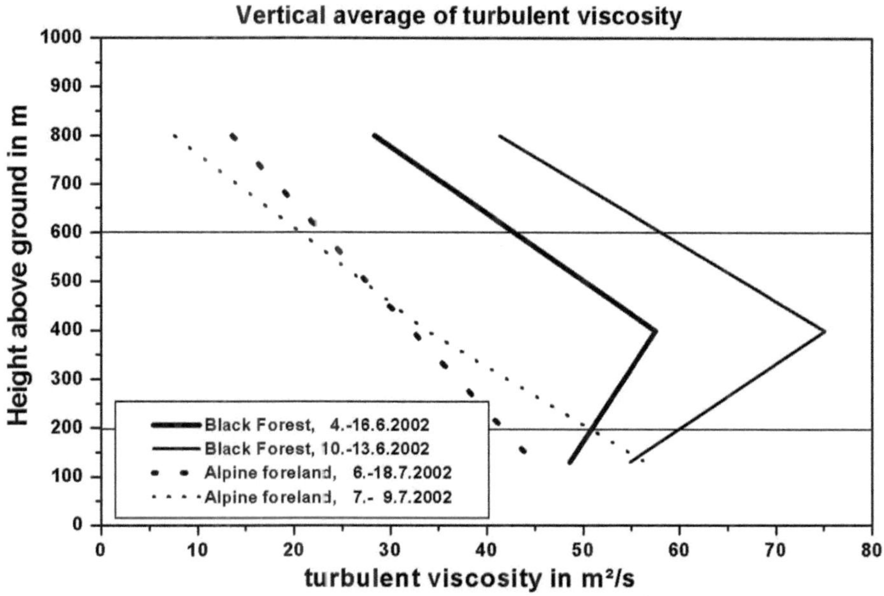

Fig. 4.31 Vertical profiles of turbulent viscosity from SODAR measurements using eq. (4.22) for flat terrain (*dashed* and *dotted lines*) and for mountainous terrain (*thin* and *thick full lines*)

clear calm nights it is also due to increased wind shear. This indicates that the overall thermal stability of the air is primarily controlling the magnitude of turbulent viscosity.

Figure 4.31 compares vertical profiles of the turbulent viscosity, which have been computed in the abovementioned manner for a flat pre-Alpine region and for mountainous terrain (the sounding has been made near a hill-top in the Black Forest in South-western Germany). For each terrain type two evaluations are presented, which look consistent. Turbulent viscosity – and thus the vertical momentum exchange – is larger over mountainous terrain than over flat terrain. Probably, small-scale secondary circulation patterns due to dynamic and thermal forcing on the hill slopes enhances the standard deviation of the vertical wind component without influencing the overall vertical wind shear so much. Therefore, according to (4.22), the turbulent viscosity above mountainous terrain turns out higher than over flat terrain.

4.4.2 Heat Flux

A method to compute the heat flux via variances from wind profiler/RASS measurements is described and compared to a derivation from an eddy-correlation technique using data from the same instruments and to airborne in-situ measurements in Bange et al. (2002). The shown examples in this paper compare well with independent measurements and modelled profiles. The variance method determines the turbulent heat

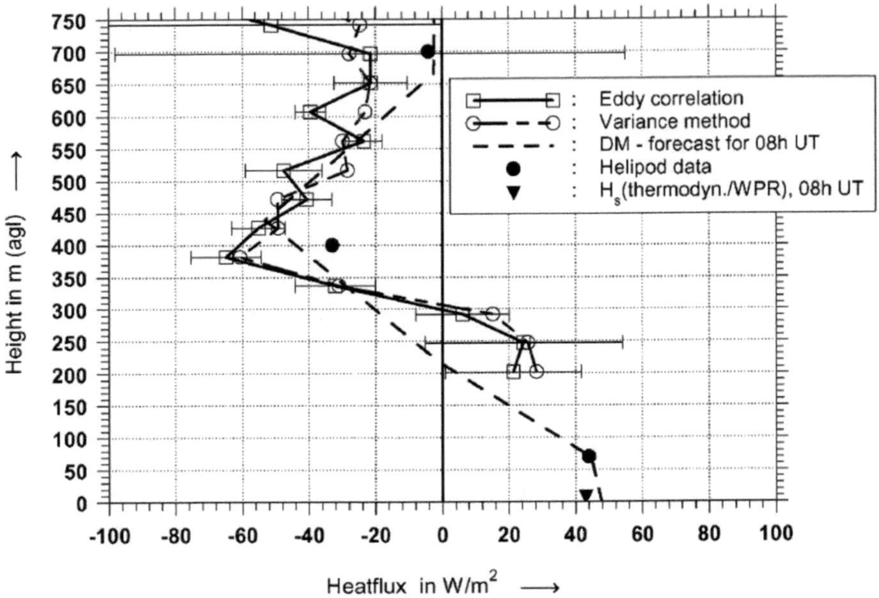

Fig. 4.32 Sensible heat flux profile (*lines* connecting squares and circles) from windprofiler/RASS measurements compared to model forecasts (*dashed line*) and in situ data (*dots* and *triangle*). From Engelbart and Bange (2002)

flux from the measured variances of the RASS sound propagation velocity (this still includes effects of the vertical wind component), of temperature, and of the vertical velocity component. Figure 4.32 shows an example.

Davis et al. (2009) determined the vertical turbulent heat flux from a vertically pointing pulsed Doppler LIDAR using an estimation of the energy dissipation rate. They started from (Davis et al. 2008):

$$\overline{w'\Theta'} \approx \frac{\Theta_0}{g}\left[\frac{\partial}{\partial z}\left(\frac{1}{2}\overline{w'^3}\right) + \frac{\varepsilon}{3}\right], \tag{4.23}$$

While the fluctuations of the vertical velocity components are taken directly from the Doppler LIDAR soundings, the dissipation ϵ can be determined in two ways. The so-called spatial spectral method estimates it from the slope of the spectrum of the longitudinal velocity correlation. The so-called temporal spectra method estimates it from the spectral energy at a given frequency and the mean horizontal wind speed.

Similarity laws for the atmospheric boundary layer may be used to determine the near-surface turbulent fluxes of sensible (*H*) as well as latent heat (*LE*) from surface-layer turbulence measurements with scintillometers. For unstable conditions, Kohsiek (1982) gives, using the structure functions for temperature, C_T and for moisture C_q:

$$H = b\left(C_T^2\right)^{3/4}, \tag{4.24}$$

$$\mathrm{LE} = 1.09b\left(C_T^2\right)^{1/4}\left(C_q^2\right)^{1/2}, \tag{4.25}$$

$$b = 0.55z\left(\frac{g}{T}\right)^{1/2}. \tag{4.26}$$

For large Bowen ratios when the humidity influence can be neglected, scintillometer measurements at one wavelength are sufficient and the heat flux can be derived from a relation between ${C_\mathrm{T}}^2$ and ${C_\mathrm{n}}^2$ from Wesely (1976) and from Monin-Obukhov similarity theory (Kleissl et al. 2008). A comparison of the structure parameters for temperature and humidity from scintillometer measurements with simultaneous eddy-covariance measurements is described in Beyrich et al. (2005), showing a consistent behaviour in time and deviations in the order of 20–35%.

4.4.3 Humidity or Water Vapour Flux

Vertical profiles of water vapour fluxes have been derived from ground-based remote sensing with high-resolution Doppler LIDAR instruments by, e.g. Giez et al. (1999), Bösenberg and Linné (2002). These fluxes can also be observed by the simultaneous operation of a Bragg-RASS for the wind fluctuations and a water vapour DIAL for the humidity fluctuations (Wulfmeyer 1999). Figure 4.33 shows an example. Here fluxes up to about 600 m above ground have been determined with a vertical resolution of about 70 m

Water vapour fluxes have been obtained by the operation of a Doppler LIDAR for the wind fluctuations and a DIAL for the humidity fluctuations as well (Linné et al. 2007). The eddy-correlation method was applied, and error estimates of $\pm 50\,\mathrm{W/m^2}$ for latent heat flux were found. Since the sampling error dominates the overall measurement accuracy, time intervals between 60 and 120 min were required for a reliable flux calculation, depending on wind speed. Rather large errors may occur with low wind speeds because the diurnal cycle restricts the useful interval length. The LIDAR flux profiles are well complemented by tower measurements at 50 and 90 m above ground and by area-averaged near surface fluxes from a network of micrometeorological stations. Water vapour flux profiles in the convective boundary layer exhibit different structures mainly depending on the magnitude of the entrainment flux. In situations with dry air above the boundary layer, a positive entrainment flux is observed which can even exceed the surface flux. Flux profiles which linearly increase from the surface to the top of the boundary layer are observed as well as profiles which decrease in the lower part and increase in the upper part of the boundary layer. In situations with humid air above the boundary layer, the entrainment flux is about zero in the upper part of the boundary layer and the profiles in most cases show a linear decrease (Linné et al. 2007).

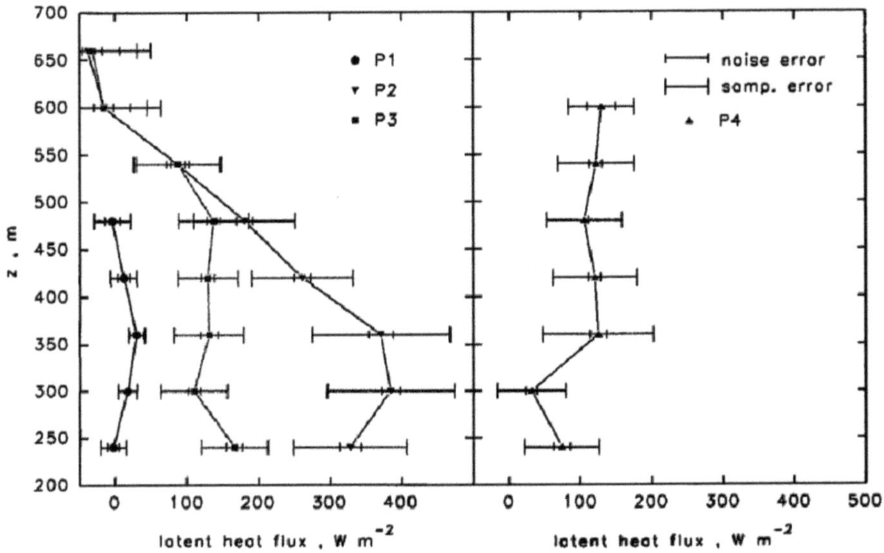

Fig. 4.33 Water vapour flux profiles from DIAL measurements (Wulfmeyer 1999)

4.5 Regional and Local Flow Systems

Up to now we have statically analyzed the vertical structure of the ABL and vertical profiles of different variables with the ABL. But there are also dynamical flow processes taking place in the ABL, caused by temporal and spatial variations of the surface conditions. Some of these regional and local flows can considerably modify again the vertical structure of the ABL. The first two subsections in this section deal with non-stationary phenomena. Low-level jets occur due to diurnal variations in the static stability of the ABL in calm weather conditions but still existing large-scale horizontal pressure gradients, while gust fronts and cold air outbreaks are boundary-layer phenomena related to severe convective events such as thunderstorms.

The second three subsections present flow features due to horizontal differences in surface properties. While land–sea differences are mainly provoked by the difference in thermal inertia of these two surface types, mountainous flow features are caused by mechanical and thermal influences.

4.5.1 Low-Level Jets

Low-level jets are a dynamical feature occurring near the top of nocturnal stable boundary layers in synoptically undisturbed weather situations. Low-level jets have already been mentioned in Section 4.3.1.2, where optical wind sounding over flat terrain was presented. More detailed measurements can often be made by acoustic

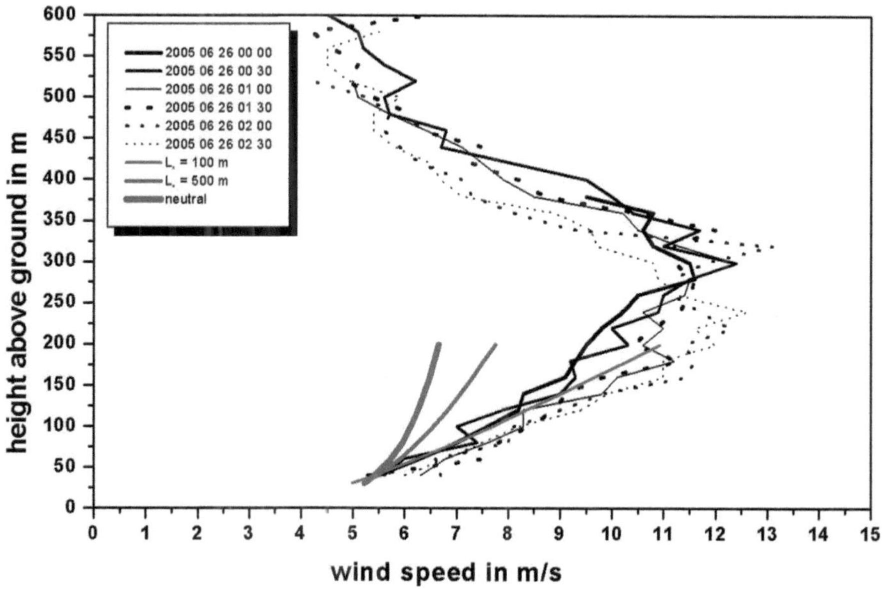

Fig. 4.34 SODAR wind profile measurements featuring a nocturnal low-level jet near Paris airport Charles de Gaulle on June 26, 2005. Six subsequent half-hourly averaged profiles are shown. Lower parts of the profiles are compared to theoretical curves (logarithmic wind profiles for different stabilities)

sounding. Here, Fig. 4.34 shows an example measured with a SODAR. The phenomenon, which must be described as an inertial oscillation due to the suddenly waning frictional force at the onset of the night due to radiative cooling of the surface, shows a wind speed maximum at about 300–350 m above ground. Such wind maxima may be relevant for pollutant dispersion as well as for air traffic security or wind energy conversion with tall wind turbines. Helmis (2007) observed low-level jets at the top of a marine boundary layer and enhanced variance of the vertical wind component with it from SODAR soundings. Further SODAR measurements of low-level jets are reported in Kallistratova et al. (2009). Empirical distributions of the maximum wind speed of these jets, the height of the jet core, and the wind turning across these jets were obtained.

Reitebuch et al. (2000) have shown that nocturnal low-level jets may be responsible for secondary nocturnal ozone concentration maxima near the ground. Their SODAR measurements (see Fig. 4.35) for the first time proved from the obtained time-height cross section of the variance of the vertical wind component, σ_w, that enhanced turbulence underneath the jet maximum takes place. This turbulence, which is produced mechanically from the wind shear, is the reason for downward turbulent transport of ozone from the residual layer towards the surface layer and causes the secondary maximum in near-surface ozone concentration.

Influences of low-level jets on the data availability from RASS instruments have already been discussed with Fig. 4.5. Missing night-time RASS temperature data above a stable nocturnal surface layer is a quite reliable indicator for the presence

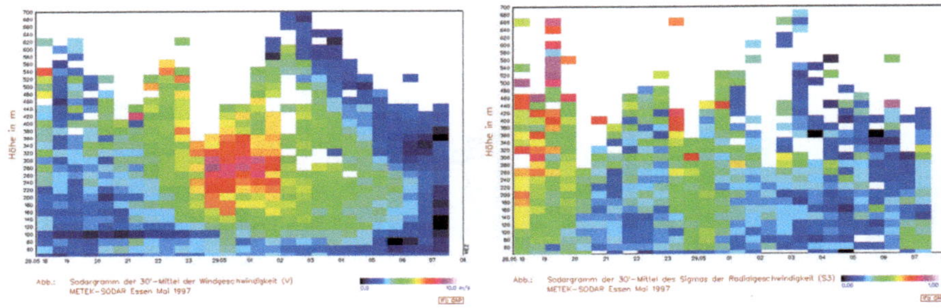

Fig. 4.35 Time-height cross sections of sodar wind measurements featuring a nocturnal low-level jet over Essen, Germany, on May 28/29, 1997. *Left*: wind speed (purple: more than 10 m s^{-1} wind speed, blue: less than 1 m s^{-1} wind speed). *Right*: Variance of the vertical velocity component as measure for the turbulence (*blue*: low values, *green*: intermediate values, *red*: high values)

of a low-level jet. Low-level jets have also coined the nocturnal wind profiles displayed in Figs. 4.15 and 4.17. Low-level jets had occurred so frequently in the spring and summer months depicted in these figures that they were detectable even from monthly mean wind profile data. Furthermore, this example is a proof that low-level jets are a regional phenomenon that develops on a horizontal scale of several hundreds of kilometres. Figures 4.15 and 4.17 show wind profiles obtained over a town which is situated quite isolated in a larger relatively flat rural area. The nocturnal urban surface layer conditions themselves (enhanced turbulence due to the high surface roughness and near-neutral to even slightly unstable thermal stratification due to the large heat capacity of the urban surface) would not have been suited for the development of low-level jets. The low-level jets rather developed over the larger rural environment of this city. The city itself is too small to eradicate the low-level jets, thus the jets are discernable from these urban wind profiles.

4.5.2 Gust Fronts and Cold Air Outflows

Gust fronts are a typical boundary layer phenomenon of larger thunderstorms (Emeis 1983). These fronts mark the leading edge of the spreading cold air underneath the thunderstorms. The boundary layer cold air pools of thunderstorms are fed by cold downdraughts and by evaporational cooling. Another source for cold air outflows is the formation of cold surface air due to nocturnal radiative cooling in mountainous terrain. This second type of cold air outflow is addressed in Sections 4.5.4 and 4.5.5 below.

Figure 4.36 shows Doppler wind LIDAR observations of an evolving thunderstorm gust front (Intrieri et al. 1990) from range-height indicator (RHI) scans. In the figure the cold air flows from the left-hand side towards the LIDAR and the line demarks the boundary of the cold air. The depicted wind arrows in this vertical cross section indicate the horizontal and vertical wind components within the plane of this cross section. The vertical flow component was derived from the quasi-horizontal wind data using the continuity equation and assuming vanishing vertical

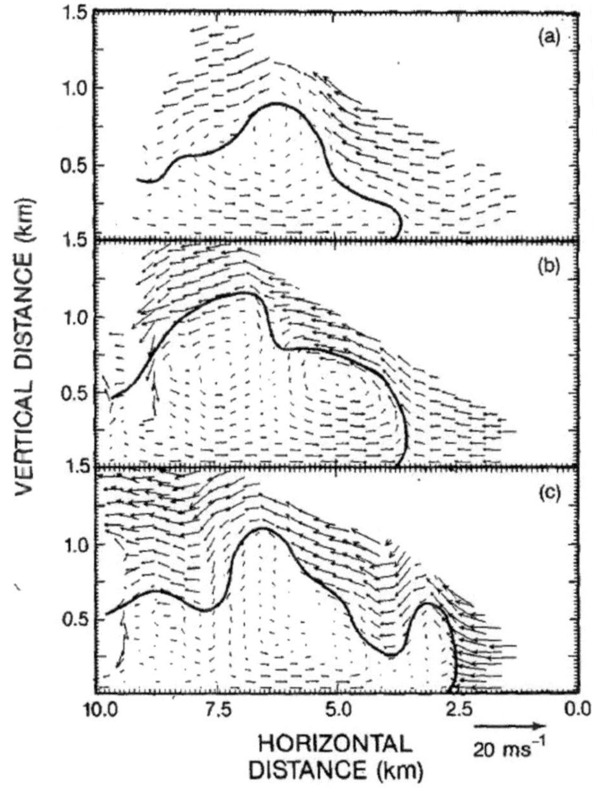

Fig. 4.36 Sequence of Doppler wind LIDAR observations of a moving thunderstorm gust front. The LIDAR was positioned at 0.0 km (Fig. 7 from Intrieri et al. 1990)

motion at the surface. The field of arrows clearly shows horizontal vortex rolls that are associated with the head of the density flow. Time-height cross sections from vertically pointing RADAR observations of thunderstorm gust fronts and outflows allow for a direct detailed analysis of vertical velocities in these outflows (Martner 1997). The maximum updraft, of almost 10 m s^{-1}, was measured in the gust front at 1.35 km above ground level and was followed by equally strong downward motion only 1 min later at a slightly higher altitude. The observations support the earlier use of the continuity method for deriving the basic pattern of vertical motions in density currents from quasi-horizontal scan data.

4.5.3 Land and Sea Breeze

Remote-sensing data from marine boundary layers are rare. This is because platforms for the installation of remote-sensing instrumentation are rare. Vertical pointing remote-sensing devices usually need fixed platforms. Deviations of a few degrees from the true vertical direction give large errors, especially if radial wind components are observed. A few experiments with SODARs on board of ships and on oil drilling platforms have been made. First texts with a wind LIDAR have been made on the FINO1 platform in the German bight (Kindler et al. 2007).

The land–sea contrast in thermal surface properties leads to the evolution of a diurnal boundary layer wind system. Due to its importance for coastal meteorology and air quality, the land–sea breeze system is a frequently studied phenomenon (Banta et al. 1993). Different remote-sensing techniques allow for the analysis of the wind speed and the aerosol distribution in the air layers involved in this circulation system. Skakalova et al. (2003) used an aerosol LIDAR to detect the different aerosol contents in the onshore and offshore flows at the Black Sea coast. While Fig. 2.9 shows the well-developed land breeze and sea breeze circulations, the results from Skakalova et al. (2003) give insight in the transition from a land breeze to a sea breeze system. They find a new layer evolving near the ground propagating inland, an old layer from the preceding land breeze regime, and an intermediate layer, which has the lowest aerosol extinction, in between the other two layers. The intermediate layer vanishes when the sea breeze is fully developed.

Banta et al. (1993) employed a Doppler LIDAR to detect the horizontal and vertical structure of the wind field of a growing sea breeze layer. Like the aerosol data from Skakalova et al. (2003), these wind data show the beginnings of the sea breeze underneath the pre-existing offshore synoptic flow at the coast and the subsequent growth of the sea breeze layer horizontally and vertically. Sometimes a dual structure to the sea breeze flow in its early formative stages was found. Initially, a shallow (less than 500 m) sea breeze formed that later became embedded in a weaker onshore flow that was about 1 km deep. Eventually, these two flows blended together to form a mature sea breeze about 1 km deep. Darby et al. (2002b) compared these Doppler LIDAR wind measurements with numerical simulations. Regional Atmospheric Modeling System (RAMS) two-dimensional simulations successfully simulated the dual structure of the sea-breeze flow when both the coastal mountain range just east of Monterey Bay and the Sierra Nevada range, peaking 300 km east of the shore, were included in the domain.

4.5.4 Flow in Mountainous Terrain, Valley, and Mountain Breeze

The most complex features of atmospheric boundary layers may be found in mountainous terrain. The ABL structure over mountainous terrain is important for local circulations, cloud formation, boundary layer-free troposphere exchange, and air quality. ABL features in mountainous terrain may be due to mechanical forcing of the flow or due to thermal forcing. Sometimes both triggering phenomena act together. We start in this subsection with mainly mechanically induced ABL features, while thermally induced features are addressed in the following subchapter. Amongst the mechanically induced phenomena are such features as valley boundary layers, channelling of the flow along the valley axis, speed up of flow over hills and crestlines, flow through gaps, or foehn flows crossing entire mountain ranges.

Figure 4.37 shows SODAR measurements of vertical profiles of the horizontal wind speed in a narrow north-south-oriented Alpine valley at the northern fringe of

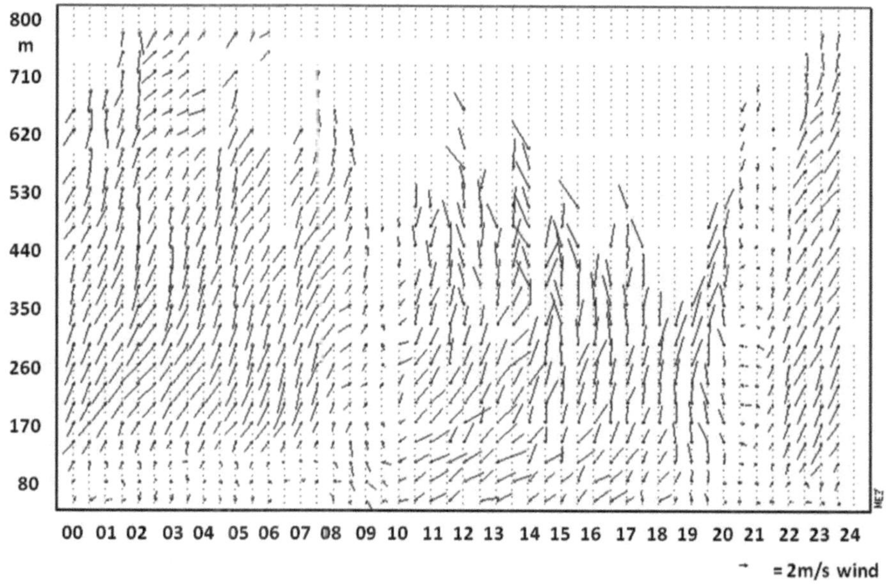

Fig. 4.37 Time-height section of the horizontal wind vector in an Alpine valley from SODAR observations. An upward pointing *arrow* designates a southerly wind, a rightward pointing a westerly wind, and so on. The length of the arrows is proportional to the wind speed. The figure shows half-hourly data for one day (from midnight to midnight) for a height interval from 50 m to 800 m above ground with a vertical resolution of 30 m

the European Alps. This plot clearly documents the diurnal variation of the channelled wind regime in such a valley during undisturbed synoptic weather conditions. At night-time out-valley (southerly) flow can be observed over the whole height range while during daytime in-valley (northerly) flow prevails. The changes between the two regimes roughly take place at 9 a.m. and 9 p.m. There is some indication that the change occurs first in lower layers. Furthermore, from the directional scatter of the wind arrows, it is obvious that the daytime flow is much more turbulent than the nocturnal flow.

While Fig. 4.37 gives an example for a vertically well-mixed convective daytime valley boundary layer, the example plotted in Fig. 4.1 presents the opposite case: a very stable valley boundary with several persisting inversions within a few hundreds of metres above the valley flow. Statistical evaluations of the SODAR soundings in Emeis et al. (2007b) show that such layering with multiple inversions occurs quite regularly in Alpine valleys. Figure 4.38 shows the frequency distribution of the height of the lowest inversion (the mixing layer height) and Fig. 4.39 shows the frequency of multiple inversions from hourly data in January. It turns out that in a wintry Alpine valley the mixing layer height is most frequent at about 100 m above ground or even lower. In 83% of all cases at least one inversion was present, and in 56% of all cases multiple inversions were observed. This multiple layering is a

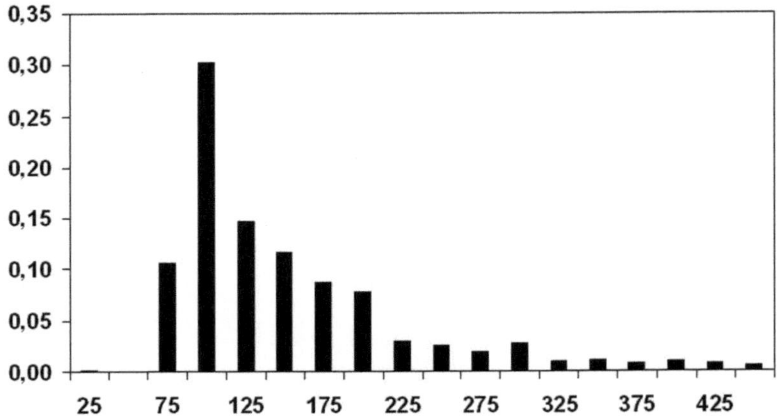

Fig. 4.38 Frequency distribution of the height of the lowest inversion (mixing layer height) in a wintry valley boundary layer from SODAR measurements (from Emeis et al. 2007b)

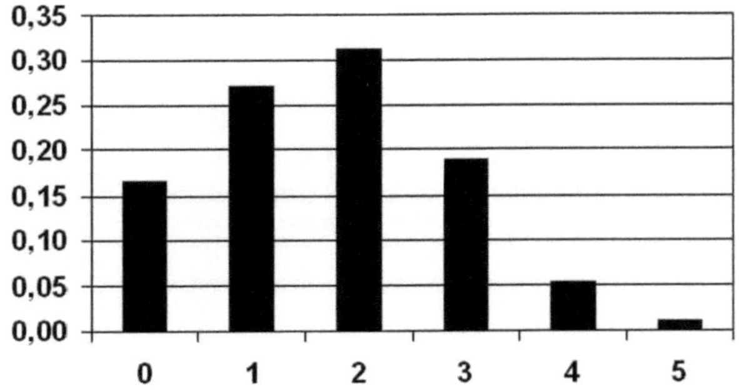

Fig. 4.39 Frequency of the occurrence of multiple inversions in a wintry valley boundary layer from SODAR measurements (from Emeis et al. 2007b)

special feature of a valley boundary layer. It originates from an interlacing of down-valley and down-slope flows. The layers are separated by temperature inversions, and each higher layer is potentially warmer than the next lower layer.

The Mistral as an example for a gap flow has already been shown in Fig. 4.18. The Mistral is a wind blowing southward through the Rhone valley towards the French Mediterranean coast. The river Rhone flows through a major gap between the Massif Central to the West and the French Alps to the East. Mechanisms responsible for the temporal evolution of the Mistral are related to the evolution of the upstream synoptic wind speed and direction conditions during the event and the upstream Froude number, calculated in the layer below the upstream inversion height. The Froude number in the case shown in Fig. 4.18 has been found to be between 0.19 and 0.75, thus systematically smaller than 1 (subcritical incident flow). With such Froude numbers, the ability for air masses to pass over the mountains is mainly governed by the comparison between the upstream inversion height and the mountain elevations (Caccia et al. 2004).

Doppler LIDARs, including HRDL and other systems with greater range, have proven capabilities to reveal the relationship between wind flow variability and terrain. For example, localized nocturnal flows emanating from valleys or canyons emptying onto a plain were studied in Colorado (Banta et al. 1995, 1996; Levinson and Banta 1995; Darby et al. 1999) and near Salt Lake City, Utah (Banta et al. 2004; Darby et al. 2006); the structure of flow within Alpine valleys was documented by Drobinski et al. (2001; 2003a,b); windstorm-type flow structure was noted in Colorado (Clark et al. 1994) and to the north of the Alps in Austria (Flamant et al. 2002; Weissmann et al. 2004); and a recurrent nocturnal LLJ structure was found in the valley of the Great Salt Lake, Utah (Banta et al. 2004). The use of Doppler LIDAR scan data, e.g. for wind energy for site assessment can replace extensive arrays of surface tower networks.

A typical phenomenon of flows over hills is the speed-up of the flow over the hill crest and a decrease of the flow speed on the downwind side of the hill. Under appropriate thermal conditions the flow can even detach from the downstream side of a hill and re-circulations can form there. SODAR measurements in Fig. 4.11 (right) demonstrate that the major part of the wind speed increase is limited to lower heights (about 50 m above the hill top). Above this height, the wind speed increase is much lower than over flat terrain (Fig. 4.11 left).

Larger north-south oriented valleys in east-west oriented mountain chains such as the European Alps are often influenced by foehn flows when the large-scale synoptic wind direction is nearly perpendicular to the mountain chain. On the northern side of the Alps, the Rhine valley between Chur (Switzerland) and the Rhine's estuary into Lake Constance near Bregenz (Austria) is such a valley. The interaction of the foehn flow with the usual valley wind systems is a complicated phenomenon. In order to get deeper insight into this interaction, the project FORM (Foehn in the Rhine Valley during MAP) was conducted in this valley in 1999. MAP was the Mesoscale Alpine Programme. An overview of FORM and some Doppler wind LIDAR measurements from this valley are given in Drobinski et al. (2007). Aerosol layers in the valley air have been monitored continuously with almost no interruption during the entire field phase of FORM with a backscatter LIDAR (Frioud et al. 2004). Vertical wind and temperature profiles were observed from five Doppler SODARs and two wind profiling RADARs (one equipped with a radio acoustic sounding system, RASS). They contributed to validate high-resolution numerical simulations of the complex interaction between the foehn and the valley flow systems (Vogt and Jaubert 2004).

4.5.5 Drainage and Katabatic Flows

Drainage and katabatic flows are purely thermally generated orographic flow features in a mountain boundary layer. Long-wave radiative loss to space leads to cooling of land, snow, and ice surfaces and a compensating downward sensible heat flux, which cools the atmospheric surface layer and forms a temperature inversion. In the presence of slopes this induces a horizontal temperature gradient equivalent to a down-slope horizontal pressure gradient force (Anderson et al. 2005, Renfrew and Anderson 2006), which drives usually shallow drainage flows. These drainage flows are often too shallow in order to be detected and monitored by ground-based remote sensing.

Drastic examples of deeper drainage flow are katabatic flows at locations of Antarctica and Greenland. The domed topography and radiative cooling of the snow surface make katabatic flows ubiquitous over these regions (Renfrew and Anderson 2006).

An autonomous Doppler SODAR wind profiling system has been operated over a 2-year period for the first time. An exemplary result from these measurements is shown in Fig. 4.40. The main results are that there is a systematic change in the

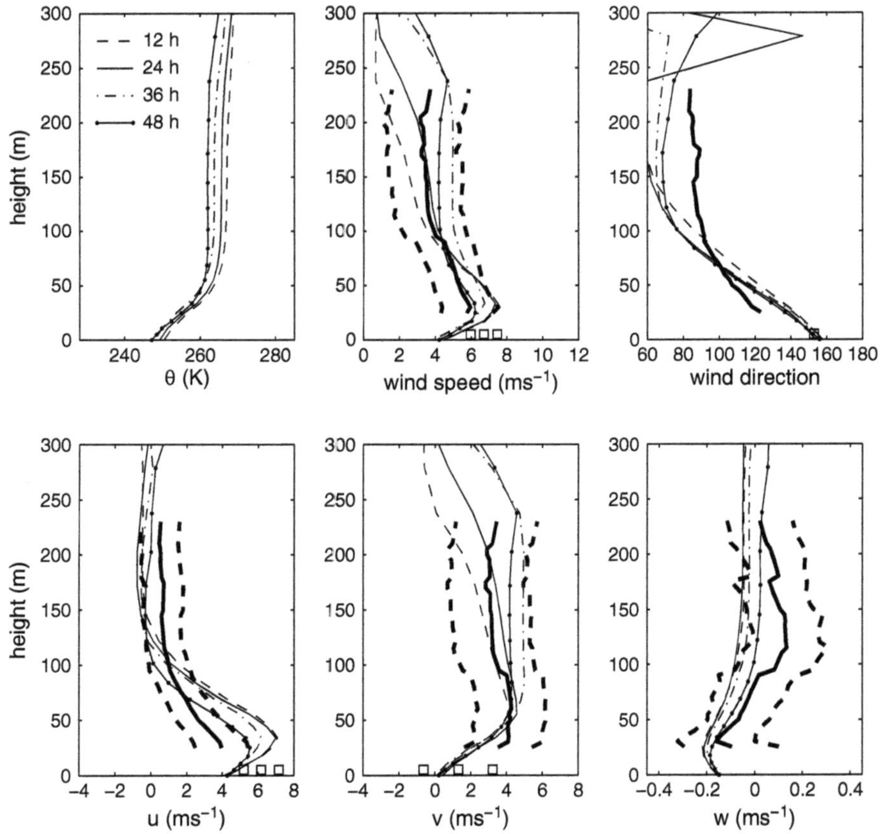

Fig. 4.40 Numerical model results (*thin lines*) and remote-sensing data (*thick lines*, *dashed lines* indicate plus/minus one standard deviation) for katabatic flow over Antarctica. (**a**) potential temperature, (**b**) wind speed, (**c**) wind direction, (**d**) downslope horizontal velocity (u), (**e**) across-slope horizontal velocity (v), and (**f**) vertical velocity (w). (Renfrew, personal communication, see also Renfrew and Anderson (2006) for further evaluations)

shape of the low-level katabatic jet with wind speed. On average, stronger katabatic flows (wind speed maxima of typically 8–10 m s^{-1}) are deeper with a jet maximum at ~20–60 m altitude; moderate katabatic flows (wind speed maxima of typically 4–8 m/s) are shallower with a jet maximum more often between 3 and 30 m. The strongest katabatic flows have a greater cross-slope component, consistent with a greater Coriolis force and frictional drag. For all katabatic wind profiles, there is a backing in wind direction with height over the lowest ~100 m, consistent with a frictional forcing at the surface. During summer there is a clear diurnal signature in the katabatic flow at all heights. It is more pronounced at 20 m and above, where the flow often ceases entirely, whereas at the surface there always appears to be some weak (2–4 m/s) katabatic flow. Such cessation events may lead to katabatic jumps, analogous to hydraulic jumps, as the flow changes from supercritical to subcritical down the slope (Renfrew and Anderson 2006).

Earlier observations of katabatic winds in Antarctica and Greenland with SODARs had already been made by Argentini and Mastrantonio (1994), Meesters et al. (1997) who also used a RASS, and Bromwich and Liu (1996). Wave structures in Antarctic katabatic flows have been analyzed from SODAR echograms by Kouznetsov (2009).

SODAR instruments can also be used to determine the seeing, a variable important for optical astronomers (Bonner et al. 2008). For such a purpose a SNODAR (see Chapter 3) has been deployed to Antarctica to monitor the height and variability of the Antarctic ABL up to heights of about 100 m above ground with 5 m vertical resolution.

4.6 Conclusions on the Applicability of Ground-Based Remote Sensing for ABL Research and Monitoring

The examples for probing the atmospheric boundary layer by ground-based remote sensing gathered in this chapter have shown the impressive progress which has been made since the last review by Wilczak et al. (1996). This progress is at least as substantial as the progress stated by Wilczak et al. (1996) in comparison to the earlier and probably first review of this subject by Derr and Little (1970). The short outlook in the next chapter will show that the development of ground-based remote sensing is still ongoing.

Acoustic remote sensing with SODAR, which had been – together with RADAR – the pioneering technology for boundary layer profiling, is still frequently used for the detection of wind and turbulence profiles and for monitoring mixing layer heights. The limited height range of the SODAR technology (which nevertheless is nearly always sufficient for the detection of shallow nocturnal mixing layers), together with possible acoustic disturbances to near-by residents due to the audible signals, are the two largest deficiencies of this otherwise cheap and robust sounding method.

Acoustic remote sensing is still alive also because it can quite easily be enhanced to a RASS by adding a pair of electromagnetic antennas to a SODAR. This allows directly detecting the temperature profile as well. The temperature profiles are valuable for the investigation of the static stability of the lower part of the ABL and can be employed to secure the assessment of the mixing-layer height in case of ambiguous acoustic information from the SODAR component of the instrument.

Optical remote sensing is presently the most rapidly developing part of the ground-based remote-sensing technology. The greatest deficiency of this technology, which used to be the rather large height of the first range gate, has nowadays been overcome by enhanced laser technology and electronics. Today, Doppler wind LIDARs which cover the height range between 20 m and 200 m are available and are increasingly used for wind energy assessments. The worldwide need for an enhanced usage of renewable energies has fostered the market for these small wind LIDARs.

In addition to the variables accessible by acoustic methods including RASS (wind, turbulence, temperature, mixing-layer height), optical remote sensing is well suited to observe aerosol, humidity, and other trace gas profiles as well. This has large relevance for precipitation, air quality, and climate impact studies. The further emergence of Raman LIDAR technology has made it possible to derive even profiles of aerosol properties. Optical and acoustic sensing technologies taken together now allow for a nearly complete probing of all major properties of the ABL.

Electromagnetic remote sensing with windprofilers has been continued in the past decades. Such profilers, which cover more or less the whole troposphere including the ABL, increasingly substitute the classical radiosonde soundings, because they can be operated in an automized mode. Like SODARs, many windprofilers have been complemented by an acoustic component to transfer them into RASS as well. By this, wind and temperature profiles can be obtained simultaneously.

Passive remote sensing with radiometers offers less vertical resolution due to the employment of weighting functions. Therefore, these instruments do not have so much relevance for genuine boundary layer research. They are used more for obtaining profiles throughout the whole troposphere and even from the stratosphere.

The development of remote sensing technologies in the past decades has led to a revival of boundary layer meteorology. In addition to the surface layer investigations by in situ instrumentation, now profiling of the whole depth of the boundary layer has become possible. This has relevance for basic studies of the dynamics and thermodynamics of the atmosphere as well as for air quality studies, assessments of renewable energy potentials, and climate change impact studies. The separated descriptions of the surface or Prandtl layer and the Ekman layer must now be combined in a more general description of the whole ABL.

References

Al-Sakka, H, Weill A, Le Gac C, Ney R, Chardenal L, Vinson J-P, Barthels L, Dupont E (2009) CURIE: a low power X-band, low atmospheric boundary layer Doppler radar. Meteorol Z 18:267–276

Anderson PS, Ladkin RS, Renfrew IA (2005) An autonomous Doppler sodar wind profiling system. J Atmos Oceanic Technol 22:1309–1325

Angevine WM, Avery SK, Ecklund WL, Carter DA (1993a) Fluxes of heat and momentum measured with a boundary layer wind profiler radar-radioacoustic sounding system. J Appl Meteor 32:73–80

Angevine WM, Avery SK, Kok JL (1993b) Virtual heat flux measurements from a boundary-layer profiler-RASS compared to aircraft measurements. J Appl Meteor 32:1901–1907

Angevine W, White AB, Avery SK (1994) Boundary layer depth and entrainment zone characterization with a boundary layer profiler. Bound-Layer Meteorol 68:375–385

Argentini S, Mastrantonio G (1994) Barrier winds recorded during two summer Antarctic campaigns and their interaction with the katabatic flows as observed by a tri-axial Doppler sodar. Int J Rem Sens 15:455–466

Asimakopoulos DN, Helmis CG, Michopoulos J (2004) Evaluation of SODAR methods for the determination of the atmospheric boundary layer mixing height. Meteor Atmos Phys 85:85–92

Banakh VA, Smalikho IN (1997) Estimation of the turbulent energy dissipation rate from the pulsed Doppler lidar data. Atmos Oceanic Opt 10:957–965

Bange J, Beyrich F, Engelbart DAM (2002) Airborne measurements of turbulent fluxes during LITFASS-98: Comparison with ground measurements and remote sensing in a case study. Theor Appl Climatol 73:35–51

Banta RM, Olivier LD, Levinson DH (1993) Evolution of the monteray bay sea-breeze layer as observed by pulsed Doppler lidar. J Atmos Sci 50:3959–3982

Banta RM, Olivier LD, Neff WD, Levinson DH, Ruffieux D (1995) Influence of canyon-induced flows on flow and dispersion over adjacent plains. Theor Appl Climatol 52:27–42

Banta RM, Olivier LD, Gudiksen PH, Lange R (1996) Implications of small-scale flow features to modeling dispersion over complex terrain. J Appl Meteor 35:330–342

Banta RM, Newsom RK, Lundquist JK, Pichugina YL, Coulter RL, Mahrt L (2002) Nocturnal low-level jet characteristics over Kansas during CASES-99. Bound-Layer Meteor 105:221–252

Banta RM, Darby LS, Fast JD, Pinto JO, Whiteman CD, Shaw WJ, Orr BD (2004) Nocturnal low-level jet in a mountain basin complex. Part I: Evolution and implications to other flow features. J Appl Meteor 43:1348–1365

Banta RM, Pichugina YL, Brewer WA (2006) Turbulent velocity-variance profiles in the stable boundary layer generated by a nocturnal low-level jet. J Atmos Sci 63:2700–2719

Behrendt A, Reichardt J (2000) Atmospheric temperature profiling in the presence of clouds with a pure rotational Raman lidar by use of an interference-filter-based polychromator. Appl Optics 39:1372–1378

Beyrich F (1995) Mixing height estimation in the convective boundary layer using sodar data. Bound-Lay Meteorol 74:1–18

Beyrich F (1997) Mixing height estimation from sodar data – a critical discussion. Atmosph Environ 31:3941–3954

Beyrich F, Görsdorf U (1995) Composing the diurnal cycle of mixing height from simultaneous SODAR and wind profiler measurements. Bound-Layer Meteorol 76:387–394

Beyrich F, Kouznetsov RD, Leps J-P, Lüdi A, Meijninger WML, Weisensee U (2005) Structure parameters for temperature and humidity from simultaneous eddy-covariance and Scintillometer measurements. Meteorol Z 14:641–649

Bingöl F, Mann J, Foussekis D (2009) Conically scanning lidar error in complex terrain. Meteorol Z 18:189–195

Blumen W, Banta RM, Burns SP, Fritts DC, Newsom R, Poulos GS, Sun J (2001) Turbulence statistics of a Kelvin-Helmholtz billow event observed in the nighttime boundary layer during the CASES-99 field program. Dyn Atmos Oceans 34:189–204

Boers R, Spinhirne JD, Hart WD (1983) Lidar observations of the fine-scale variability of marine stratocumulus clouds. J Appl Meteorol 27:797–810

Bonner CS, Ashley MCB, Lawrence JS, Storey JWV, Luong-Van DM, Bradley SG (2008) Snodar: a new instrument to measure the height of the boundary layer on the Antarctic plateau. Proceedings of SPIE 7014: 70146I, p. 7, http://spie.org

Bösenberg J, Linné H (2002) Laser remote sensing of the planetary boundary layer. Meteorol Z 11:233–240
Bromwich DH, Liu Z (1996) An observational study of the katabatic wind confluence zone near siple coast, West Antarctica. Mon Wea Rev 124:462–477
Brooks IM (2003) Finding boundary layer top: Application of a wavelet covariance transform to lidar backscatter profiles. J Atmos Oceanic Technol 20:1092–1105
Browning KA, Wexler R (1968) The determination of kinematic properties of a wind field using Doppler radar. J Appl Meteorol 7:105–108
Caccia J-L, Guénard V, Benech B, Campistron B, Drobinski P (2004) Vertical velocity and turbulence aspects during mistral events as observed by UHF wind profilers. Annales Geophysicae 22:3927–3936
Campistron B, Huggins AW, LONG AB (1991) Investigations of a winter mountain storm in Utah. Part III: Single-Doppler radar measurements of turbulence. J Atmos Sci 48:1306–1318
Camy-Peyret C, Bergovist B, Galle B, Carleer M, Clerbaux C, Colin R, Fayt C, Goutail F, Nunes-Pinharanda M, Pommereau JP, Hausmann M, Platt U, Pundt I, Rudolph T, Hermans C, Simon PC, Vandaele AC, Plane JMC, Smith N (1996) Intercomparison of instruments for tropospheric measurements using differential optical absorption spectroscopy. J Atmos Chem 23:51–80
Carter DA, Gage KS, Ecklund WL, Angevine WM, Johnston PE, Riddle AC, Wilson J, Williams CR (1995) Developments in UHF lower tropospheric wind profiling at NOAA's aeronomy laboratory. Radio Sci 30:977–1001
Clark TL, Hall WD, Banta RM (1994) Two- and three-dimensional simulations of the 9 January 1989 windstorm: Comparison with observations. J Atmos Sci 51:2317–2343
Cohn SA, Angevine WM (2000) Boundary layer height and entrainment zone thickness measured by lidars and wind-profiling radars. J Appl Meteorol 39:1233–1247
Collier CG, Davies F, Bozier KE, Holt AR, Middleton DR, Pearson GN, Siemen S, Willets DV, Upton GJG, Young RI (2005) Dual-Doppler lidar measurements for improving dispersion models. Bull Amer Meteor Soc 86:825–838
Contini D, Cava D, Martano P, Donateo A, Grasso FM (2009) Comparison of indirect methods for the estimation of boundary layer height over flat-terrain in a coastal site. Meteorol Z 18: 309–320
Cooney J (1972) Measurement of atmospheric temperature profiles by Raman backscatter. J Appl Meteorol 11:108–112
Crewell S, Mech M, Reinhardt T, Selbach C, Betz HD, Brocard E, Dick G, O'Connor E, Fischer J, Hanisch T, Hauf T, Hünerbein A, Delobbe L, Mathes A, Peters G, Wernli H, Wiegner M, Wulfmeyer V (2008) The general observation period 2007 within the priority program on quantitative precipitation forecasting: Concept and first results. Meteorol Z 17:849–866
Darby LS, Allwine KJ, Banta RM (2006) Nocturnal low-level jet in a mountain basin complex. Part II: Transport and diffusion of tracer under stable conditions. J Appl Meteor 45:740–753
Darby LS, Banta RM, Brewer WA, Neff WD, Marchbanks RD, McCarty BJ, Senff CJ, White AB, Angevine WM, Williams EJ (2002a) Vertical variations in O_3 concentrations before and after a gust front passage. J Geophys Res 107:4321. doi: 10.1029/2001JD000996
Darby LS, Banta RM, Pielke Sr RA (2002b) Comparisons between mesoscale model terrain sensitivity studies and Doppler lidar measurements of the sea breeze at Monterey Bay. Mon Wea Rev 130:2813–2838
Darby LS, Neff WD, Banta RM (1999) Multiscale analysis of a meso-frontal passage in the complex terrain of the Colorado Front Range. Mon Wea Rev 127:2062–2081
Davis JC, Bozier KE, Collier CG, Davies F (2008) Spatial variations of sensible heat flux over an urban area. Meteor Appl 15:367–380
Davis JC, Collier CG, Davies F, Pearson GN, Burton R, Russell A (2009) Doppler lidar observations of sensible heat flux and intercomparisons with a ground-based energy balance station and WRF model output. Meteorol Z 18:155–162
Davis KJ, Gamage N, Hagelberg CR, Kiemle C, Lenschow DH, Sullivan PP (2000) An objective method for deriving atmospheric structure from airborne lidar observations. J Atmos Oceanic Technol 17:1455–1468

de Haij M, Wauben W, Klein Baltink H (2006) Determination of mixing layer height from ceilometer backscatter profiles. In: Slusser JR, Schäfer K, Comerón A (eds) Remote sensing of clouds and the atmosphere XI. Proceedings of SPIE Vol 6362: paper 63620R, http://spie.org

Denmead OT (2008) Approaches to measuring fluxes of methane and nitrous oxide between landscapes and the atmosphere. Plant and Soil 309:5–24

Derr VE, Little CG (1970) A comparison of remote sensing of the clear atmosphere by optical, radio, and acoustic radar techniques. Appl Opt 9:1976–1992

Devara PCS, Ernest Ray P, Murthy BS, Pandithurai G, Sharma S, Vernekar KG (1995) Intercomparison of nocturnal lower-atmospheric structure observed with LIDAR and SODAR Techniques at Pune, India. J Appl Meteorol 34:1375–1383

Drobinski P, Dabas AM, Haeberli C, Flamant PH (2001) On the small-scale dynamics of flow splitting in the Rhine Valley during a shallow foehn event. Bound-Lay Meteorol 99:277–296

Drobinski P, Dabas AM, Haeberli C, Flamant PH (2003a) Statistical characterization of the flow structure in the Rhine Valley. Bound-Lay Meteorol 106:483–505

Drobinski P, Haeberli C, Richard E, Lothon M, Dabas AM, Flamant PH, Furger M, Steinacker R (2003b) Scale interaction processes during MAP-IOP 12 South Foehn Event in the Rhine Valley. Quart J Roy Meteorol Soc 129:729–754

Drobinski P, Steinacker R, Richner H, Baumann-Stanzer K, Beffrey G, Benech B, Berger H, Chimani B, Dabas A, Dorninger M, Dürr B, Flamant C, Frioud M, Furger M, Gröhn I, Gubser S, Gutermann T, Häberli C, Häller-Scharnhost E, Jaubert G, Lothon M, Mitev V, Pechinger U, Piringer M, Ratheiser M, Ruffieux D, Seiz G, Spatzierer M, Tschannett S, Vogt S, Werner R, Zängl G (2007) Föhn in the Rhine Valley during MAP: A review of its multiscale dynamics in complex valley geometry. Quart J Roy Meteor Soc 133:897–916

Dupont E, Menut L, Carissimo B, Pelon J, Flamant P (1999) Comparison between the atmospheric boundary layer in Paris and its rural suburbs during the ECLAP experiment. Atmos Environ 33:979–994

Eberhard WL, Cupp RE, Healey KR (1989) Doppler lidar measurement of profiles of turbulence and momentum flux. J Atmos Oceanic Technol 6:809–819

Ecklund WL, Carter DA, Balsley BB (1988) A UHF Wind Profiler for the Boundary Layer: Brief Description and Initial Results. J Atmos Oceanic Technol 5:432–441

Edson JB, Zappa CJ, Ware JA, McGillis WR, Hare JE (2004) Scalar flux profile relationships over the open ocean. J Geophys Res 109, C08S09, doi:10.1029/2003JC001960

Emeis S (1983) Development of a thunderstorm complex. Contr Phys Atmosph 56:399–404

Emeis S (2001) Vertical variation of frequency distributions of wind speed in and above the surface layer observed by sodar. Meteorol Z 10:141–149

Emeis S (2004) Parameterization of turbulent viscosity over orography. Meteorol Z 13:33–38

Emeis S (2008) Examples for the determination of turbulent (sub-synoptic) fluxes with inverse methods. Meteorol Z 17:3–11

Emeis S, Schäfer K (2006) Remote sensing methods to investigate boundary-layer structures relevant to air pollution in cities. Bound-Lay Meteorol 121:377–385

Emeis S, Türk M (2004) Frequency distributions of the mixing height over an urban area from SODAR data. Meteorol Z 13:361–367

Emeis S, Frank HP, Fiedler F (1995) Modification of air flow over an escarpment – results from the Hjardemaal experiment. Bound-Lay Meteorol 74:131–161

Emeis S, Münkel C, Vogt S, Müller WJ, Schäfer K (2004) Atmospheric boundary-layer structure from simultaneous SODAR, RASS, and ceilometer measurements. Atmos Environ 38: 273–286

Emeis S, Jahn C, Münkel C, Münsterer C, Schäfer K (2007a) Multiple atmospheric layering and mixing-layer height in the Inn valley observed by remote sensing. Meteorol Z 16:415–424

Emeis S, Baumann-Stanzer K, Piringer M, Kallistratova MA, Kouznetsov R, Yushkov V (2007b) Wind and turbulence in the urban boundary layer – analysis from acoustic remote sensing data and fit to analytical relations. Meteorol Z 16:393–406

Emeis S, Harris M, Banta RM (2007c) Boundary-layer anemometry by optical remote sensing for wind energy applications. Meteorol Z 16:337–347

Emeis S, Schäfer K, Münkel C (2008) Surface-based remote sensing of the mixing-layer height – a review. Meteorol Z 17:621–630

Emeis S, Schäfer K, Münkel C (2009) Observation of the structure of the urban boundary layer with different ceilometers and validation by RASS data. Meteorol Z 18:149–154

Engelbart D (1998) Determination of boundary-layer parameters using wind profiler/RASS and sodar/RASS. Proceedings of the 9th ISARS, Vienna, 192–195

Engelbart D (2005) Bodengebundene fernerkundung am meteorologischen observatorium lindenberg. Promet 31:134–147

Engelbart DAM, Bange J (2002) Determination of boundary-layer parameters using wind profiler/RASS and sodar/RASS in the frame of the LITFASS project. Theor Appl Climatol 73:53–65

Engelbart D, Klein Baltink H (1997) Heat flux measurements by wind profiler/RASS – a comparison with tower data. Proc COST-76 Profiler Workshop, Engelberg, Switzerland, paper 7–10

Engelbart D, Steinhagen H, Kallistratova M (2002) LINEX-2000: Assessment of different methods for determination of reference flux profiles. Proceedings of the 11th ISARS, Rome, 339–342

Engelbart DAM, Kallistratova MA, Kouznetsov R (2007) Determination of the turbulent fluxes of heat and momentum in the ABL by ground-based remote sensing techniques (a Review). Meteorol Z 16:325–335

Eresmaa N, Karppinen A, Joffre SM, Räsänen J, Talvitie H(2006) Mixing height determination by ceilometer. Atmos Chem Phys 6:1485–1493 available from www.atmos-chem-phys.net/6/1485/2006/

Feltz WF, Smith WL, Howell HB, Knuteson RO, Woolf H, Revercomb HE (2003) Near-continuous profiling of temperature, moisture, and atmospheric stability using the atmospheric emitted radiance interferometer (AERI). J Appl Meteorol 42:584–597

Flamant C, Pelon J, Flamant PH, Durand P (1997) Lidar determination of the entrainement zone thickness at the top of the unstable marin atmospheric boundary-layer. Bound-Layer Meteorol 83:247–284

Flamant C, Drobinski P, Nance L, Banta R, Darby L, Dusek J, Hardesty M, Pelon J, Richard E (2002) Gap flow in an Alpine valley during a shallow south foehn event: Observations, numerical simulations and hydraulic analog. Quart J Roy Meteor Soc 128:1173–1210

Flores E, Schäfer K, Black J, Harig R, Jahn C (2007) Remote sensing of aircraft exhaust temperature and composition by passive Fourier Transform Infrared (FTIR). Proceedings of SPIE 6745:67451Q, http://spie.org

Frehlich R, Cornman L (2002) Estimating spatial velocity statistics with coherent Doppler lidar. J Atmos Oceanic Technol 19:355–366

Frioud M, Mitev V, Matthey R, Richner H, Furger M, Gubser S (2004) Variation of the aerosol stratification over the Rhine valley during föhn development: a backscatter lidar study. Meteorol Z 13:175–181

Fritts DC, Nappo C, Riggin DM, Balsley BB, Eichinger WE, Newsom RK (2003) Analysis of ducted motions in the stable nocturnal boundary layer during CASES-99. J Atmos Sci 60: 2450–2472

Gal-Chen T, Mei X, Eberhard WL (1992) Estimation of atmospheric boundary layer fluxes and other turbulence parameters from Doppler LIDAR data. J Geophys Res 97:18409–18423

Giez A, Ehret G, Schwiezow RL, Davis KJ, Lenschow DH (1999) Water vapor flux measurements from ground-based vertically pointed water vapor differential absorption and Doppler lidars. J Atmos Oceanic Technol 16:237–250

Grimsdell AW, Angevine WM (1998) Convective boundary layer height measurement with wind profilers and comparison to cloud base. J Atmos Oceanic Technol 15:1331–1338

Grund CJ, Banta RM, George JL, Howell JN, Post MJ, Richter RA, Weickmann AM (2001) High-resolution Doppler lidar for boundary layer and cloud research. J Atmos Oceanic Technol 18:376–393

Hair JW, Caldwell LM, Krueger DA, She CY (2001) High-spectral-resolution lidar with iodine-vapor filters: measurement of atmospheric-state and aerosol profiles. Appl Optics 40: 5280–5294

Hayden KL, Anlauf KG, Hoff RM, Strapp JW, Bottenheim JW, Wiebe HA, Froude FA, Martin JB, Steyn DG, McKendry IG (1997) The vertical chemical and meteorological structure of the boundary layer in the Lower Fraser valley during Pacific '93. J Atmos Environ 31: 2089–2105

Helmis CG (2007) An experimental case study of the mean and turbulent characteristics of the vertical structure of the atmospheric boundary layer over the sea. Meteorol Z 16: 375–381

Helmis CG, Wang Q, Halios CH, Wang S, Sgouros G (2004) On the vertical turbulent structure of the marine atmospheric boundary layer. Sixteenth Symposium on Boundary Layers, Portland, paper 11.1

Hennemuth B, Kirtzel H-J (2008) Towards operational determination of boundary layer height using sodar/RASS soundings and surface heat flux data. Meteorol Z 17:283–296

Hirsch L, Peters G (1998) Abilities and limitation of a radar-RASS wind profiler for the measurement of momentum flux in the planetary boundary layer. Meteorol Z NF 7:336–344

Hooper WP, Eloranta E (1986) Lidar measurements of wind in the planetary boundary layer: the method, accuracy and results from joint measurements with radiosonde and kytoon. J Clim Appl Meteorol 25:990–1001

Intrieri JM, Bedard AJ Jr, Hardesty RM (1990) Details of colliding thunderstorm outflows as observed by Doppler lidar. J Atmos Sci 47:1081–1098

Kallistratova MA, Kouznetsov RD, Kuznetsov DD, Kuznetsova IN, Nakhaev M, Chirokova G (2009) Summertime low-level jet characteristics measured by sodar over rural and urban areas. Meteorol Z 18:289–295

Kelley N, Shirazi M, Jager D, Wilde S, Adams J, Buhl M, Sullivan P, Patton E (2004) Lamar low-level jet project interim report. NREL/TP-500-34593. Golden, CO, National Renewable Energy Laboratory, p. 216

Kelley ND, Jonkman BJ, Scott GN, Bialasiewicz JT, Redmond LS (2005) The impact of coherent turbulence on wind turbine aeroelastic response and its simulation. Windpower, 15–18 May 2005, Denver, Colorado and NREL Report NREL/CP-500-38074. Golden, CO, National Renewable Energy Laboratory

Kindler D, Oldroyd A, MacAskill A, Finch D (2007) An eight month test of the Qinetiq ZephIR system: Preliminary results. Meteorol Z 16:479–489

Kleissl J, Gomez J, Hong S-H, Hendrickx JMH, Rahn T, Defoor WL (2008) Large aperture Scintillometer intercomparison study. Bound-Lay Meteorol 128:133–150

Kohsiek W (1982) Measuring CT2, Cq2 und CTq in the unstable surface layer, and relations to the vertical fluxes of heat and moisture. Bound-Lay Meteorol 24: 89–107

Kouznetsoc RD (2009) The summertime ABL structure over an Antarctic oasis with a vertical Doppler sodar. Meteorol Z 18:163–167

Kouznetsov RD, Kramar VF, Beyrich F, Engelbart D (2004) Sodar-based estimation of TKE and momentum flux profiles in the atmospheric boundary layer: Test of a parameterization model. Meteor Atmos Phys 85:93–99

Kouznetsov R, Kramar VF, Kallistratova MA (2007) The vertical structure of turbulent momentum flux in the lower part of the atmospheric boundary layer. Meteorol Z 16:367–373

Kropfli RA (1986) Single Doppler radar measurements of turbulence profiles in the convective boundary layer. J Atmos Oceanic Technol 3:305–313

Lammert A, Bösenberg J (2006) Determination of the convective boundary-layer height with laser remote sensing. Bound-Lay Meteorol 119:159–170

Lenschow DH, Wulfmeyer VO, Senff CJ (2000) Measuring second- through fourth-order moments in noisy data. J Atmos Oceanic Technol 17:1330–1347

Levinson DH (1998) Evaluation of Doppler lidar measurements of momentum flux and wind variability along an upwelling coast. NOAA Tech Memo ERL ETL-294, p. 105

Levinson DH, Banta RM (1995) Observations of a terrain-forced mesoscale vortex and canyon drainage flows along the front range of the Colorado rockies. Mon Wea Rev 123: 2029–2050

Linné H, Hennemuth B, Bösenberg J, Ertel K (2007) Water vapour flux profiles in the convective boundary layer. Theor Appl Climatol 87:201–211

Lippmann J, Bauer M, J, Peters J (1996) Methods of virtual heat flux determination from boundary later wind profiler/RASS measurements. Contr Phys Atmos 69:119–128

Löhnert U, Turner DD, Crewell S (2009) Ground-based temperature and humidity profiling using spectral infrared and microwave observations. Part I: Simulated retrieval performance in clear-sky conditions. J Appl Meteor Climatol 48:1017–1032

Lokoshchenko MA (2002) Long-term sodar observations in Moscow and a new approach to potential mixing determination by radiosonde Data. J Atmos Oceanic Technol 19:1151–1162

Lokoshchenko MA, Perepelkin VG, Semenova NV (2007) Standard deviation of the wind vertical component and its dynamics in Moscow by the sodar data Meteorol Z 16:407–414

Lokoshchenko MA, Yavlyaeva EA, Kirtzel H-J (2009) Sodar data about wind profiles in Moscow city. Meteorol Z 18:321–330

Lothon M, Lenschow DH, Mayor SD (2006) Coherence and scale of vertical velocity in the convective boundary layer from a Doppler lidar. Bound-Lay Meteor 121:521–536

Malicet J, Daumont D, Charbonnier J, Parisse C, Chakir A, Brion J (1995) Ozone UV spectroscopy. II. Absorption cross-sections and temperature dependence. J Atmos Chem 21:263–273

Martner BE (1997) Vertical velocities in a thunderstorm gust front and outflow. J Appl Meteor 36:615–622

Martucci G, Srivastava MK, Mitev V, Matthey R, Frioud M, Richner H (2004) Comparison of lidar methods to determine the Aerosol Mixed Layer top. In: Schäfer K, Comeron A, Carleer M, Picard RH (eds), Remote sensing of clouds and the atmosphere VIII. Proceedings of SPIE 5235:447–456, http://spie.org

Mattis I, Ansmann A, Althausen D, Jaenisch V, Wandinger U, Müller D, Arshinov YF, Bobrovnikov SM, Serikov IB (2002) Relative-humidity profiling in the troposphere with a Raman lidar. Appl Opt 41:6451–6462

Maughan RA, Spanton AM, Williams ML (1982) An analysis of the frequency distribution of SODAR derived mixing heights classified by atmospheric stability. Atmos Environ 16:1209–1218

Meesters AGCA, Blink NJ, Henneken EAC, Vugts HF, Cannemeijer F (1997) Katabatic wind profiles over the Greenland ice sheet: observation and modelling. Bound-Lay Meteorol 85:475–496

Melfi SH, Spinhirne JD, Chou SH, Palm SP (1985) Lidar observation of the vertically organized convection in the planetary boundary layer over the ocean. J Clim Appl Meteorol 24:806–821

Melling H, List R (1980) Characteristics of vertical velocity fluctuations in a convective urban boundary layer. J Appl Meteorol 19:1184–1195

Menut L, Flamant C, Pelon J, Flamant PH (1999) Urban boundary-layer height determination from Lidar measurements over the Paris area. Appl Opt 38:945–954

Middleton DR, Davies F (2004) Evaluation of dispersion model parameters by dual Doppler lidar over west London, UK. 9th International Conference on Harmonization, paper 5.25

Münkel C (2007) Mixing height determination with lidar ceilometers – results from Helsinki testbed. Meteorol Z 16:451–459

Münkel C, Räsänen J (2004) New optical concept for commercial lidar ceilometers scanning the boundary layer. Proceedings of SPIE Vol 5571:364–374, http://spie.org

Nedeljkovic D, Hauchecorne A, Chanin M-L (1993) Rotational Raman lidar to measure the atmospheric temperature from ground to 30 km. IEEE Trans Geosci Remote Sens 31:90–101

Neff WD, Coulter RL (1986) Acoustic remote sensing. In: Lenschow DH (ed.) probing the atmospheric boundary layer. American Meteorological Society, Boston, MA, 201–239

Newsom RK, Banta RM (2003) Shear-flow instability in the stable nocturnal boundary layer as observed by Doppler lidar during CASES-99. J Atmos Sci 30:16–33

Newsom RK, Ligon D, Calhoun RJ, Heap R, Cregan E, Princevac M (2005) Retrieval of microscale wind and temperature fields from single- and dual-Doppler lidar data. J Appl Meteor 44:1324–1345

Pekour MS, Kallistratova MA (1993) Sodar study of the boundary layer over Moscow for air pollution application. Appl Phys B 57:49–55

Pekour MS, Kallistratova MA, Lokoshchenko MA, Petenko IV (1993) Acoustic sounding study of the mixing layer over a city. In: Optical monitoring of the environment. Proceedings of SPIE 2107:169–193, http://spie.org

Peña A, Gryning S-E, Hasager CB (2009) Comparing mixing-length models of the diabatic wind profile over homogeneous terrain. Theor Appl Climatol 100:325–335

Pérez IA, Sánchez ML, García MÁ, de Torre B (2009) Boundary layer structure and stability classification validated with CO2 concentrations over the Northern Spanish Plateau. Ann Geophys 27:339–349

Peters G, Kirtzel HJ(1994) Measurements of momentum flux in the boundary layer by RASS. J Atmos Ocean Technol 11:63–75

Peters G, Hinzpeter H, Baumann G (1985) Measurements of heat flux in the atmospheric boundary layer by sodar and RASS: a first attempt. Radio Sci 20:1555–1564

Peters G, Fischer B, Andersson T (2002) Rain observations with a vertically looking Micro Rain Radar (MRR). Boreal Env Res 7 353–362

Peters G, Fischer B, Münster H, Clemens M, Wagner A (2005) Profiles of raindrop size distributions as retrieved by microrain radars. J Appl Meteorol 44:1930–1949

Philbrick CR (2002) Raman lidar descriptions of lower atmosphere processes. Proceedings of the 21st ILRC, Valcartier, Quebec Canada:535–545

Philbrick CR, Mulik KR (2001) Application of Raman lidar to air quality measurements. Proceedings of SPIE 4035:22–33, http://spie.org

Pichugina YL, Banta RM, Kelley ND, Sandberg SP, Machol JL, Brewer WA (2004) Nocturnal low-level jet characteristics over southern Colorado. 16th Symposium on Boundary Layers and Turbulence, 9–13 August, 2004, Portland, Maine

Pichugina YL, Banta RM, Kelley ND (2005a) Application of high-resolution Doppler lidar data for wind energy assessment. Paper 6.4 (CD), 2nd Symposium on Lidar Atmospheric Applications. 85th AMS Annual Meeting, 9–13 January 2005, San Diego, CA

Pichugina YL, Banta RM, Brewer WA (2005b) Vertical profiles of velocity variances and TKE using Doppler-lidar scan data. 17th Symposium on Boundary Layers and Turbulance 22–26 May 2006 San Diego, paper 7.1

Piironen AK, Eloranta EW (1995) Convective boundary layer depths and cloud geometrical properties obtained from volume imaging lidar data. J Geophys Res 100:25569–25576

Platt U, Heue K-P, Pöhler D (2009) Two- and three-dimensional observation of trace gas and aerosol distributions by DOAS Techniques. In: Kim YJ, Platt U, Gu MB, Iwahashi H (eds) Atmospheric and Biological environmental monitoring, 1. Springer Netherlands, 3–11

Pospichal B, Crewell S (2007) Boundary layer observations in west Africa using a novel microwave radiometer. Meteorol Z 16:513–523

Potvin G, Rogers RR (1999) Measuring vertical heat flux with RASS. Meteor Atmos Phys 71: 91–103

Potvin G, Macpherson JI, Rogers RR, Donaldson NR, Strapp JW (1998) Analysis of heat flux by RASS and comparison with airplane measurements. Meteorol Z NF 7:262–270

Poulos GS, Blumen W, Fritts DC, Lundquist JK , Sun J, Burns S, Nappo C, Banta RM, Newsom RK, Cuxart J, Terradellas E, Balsley B, Jensen M (2002) CASES-99: A comprehensive investigation of the stable nocturnal boundary layer. Bull Amer Meteorol Soc 83: 555–581

Puygrenier V, Bénech B, Campistron B, Lohou F, Said F, Brut A, Moppert C, Dupont E (2004) Determination of turbulent parameters in the atmospheric boundary layer with an UHF wind profiler. Comparison with in-situ measurements. 16th Symposium on Boundary Layers and Turbulance, Portland, paper 11.1

Rabin RM, Doviak RJ, Sundara-Rajan A (1982) Doppler radar observations of momentum flux in a cloud-less convective layer with rolls. J Atmos Sci 39:851–863

Radlach M, Behrendt A, Wulfmeyer V (2008) Scanning rotational Raman lidar at 355 nm for the measurement of tropospheric temperature fields. Atmos Chem Phys 8:159–169

Reitebuch O, Straßburger A, Emeis S, Kuttler W (2000) Nocturnal secondary ozone concentration maxima analysed by SODAR observations and surface measurements. Atmos Environ 34: 4315–4329

Renfrew IA, Anderson PS (2006) Profiles of katabatic flow in summer and winter over Coats land, Antarctica. Quart J Roy Meteor Soc 132:779–802

Rose T, Crewell S, Löhnert U, Simmer C (2005) A network suitable microwave radiometer for operational monitoring of the cloudy atmosphere. Atmos Res 75:183–200

Ruffieux D, Nash J, Jeannet P, Agnew JL (2006) The COST 720 temperature, humidity, and cloud profiling campaign: TUC. Meteorol Z 15:5–10

Schäfer K, Haus R, Heland J, Haak A (1995) Measurements of atmospheric trace gases by emission and absorption spectroscopy with FTIR. Ber Bunsen Gesell 99:405–411

Schäfer K, Emeis SM, Rauch A, Münkel C, Vogt S (2004) Determination of mixing-layer heights from ceilometer data. In: Schäfer K, Comeron AT, Carleer MR, Picard RH, Sifakis N (eds): Remote sensing of clouds and the atmosphere IX. Proceedings of SPIE 5571: 248–259, http://spie.org

Schäfer K, Emeis S, Junkermann W, Münkel C (2005) Evaluation of mixing layer height monitoring by ceilometer with SODAR and microlight aircraft measurements. In: Schäfer K, Comeron AT, Slusser JR, Picard RH, Carleer MR, Sifakis N (eds): Remote sensing of clouds and the atmosphere X. Proceedings of SPIE 5979:59791I-1–59791I-11, http://spie.org

Schäfer K, Emeis S, Hoffmann H, Jahn C (2006a) Influence of mixing layer height upon air pollution in urban and sub-urban areas. Meteorol Z 15:647–658

Schäfer K, Hoffmann H, Emeis S, Wittig J, Vergeiner J (2006b) Highway emission study by DOAS within the Inn valley near Innsbruck. Proceedings of SPIE 6362: 63621D, http://spie.org

Schneider JM, Lilly DK (1999) An observational and numerical study of a sheared convective boundary layer. Part I: Phoenix II observations, statistical description, and visualization. J Atmos Sci 56:3059–3078

Schneider M, Hase F (2009) Ground-based FTIR water vapour profile analyses. Atmos Meas Tech 2:609–619

Seibert P, Beyrich F, Gryning S-E, Joffre S, Rasmussen A, Tercier P (2000) Review and intercomparison of operational methods for the determination of the mixing height. Atmos Environ 34:1001–1027

Senff C, Bösenberg J, Peters G, Schaberl T (1996) Remote sensing of turbulent ozone fluxes and the ozone budget in the convective boundary layer with DIAL and Radar-RASS: A case study. Contrib Atmos Phys 69:161–176

Sicard M, Pérez C, Rocadenbosch F, Baldasano JM, García-Vizcaino D (2006) Mixed-layer depth determination in the Barcelona coastal area from regular lidar measurements: Methods, results and limitations. Bound-Lay Meteorol 119:135–157

Skakalova TS, Savov PB, Grigorov IV, Kolev IN (2003) Lidar observation of breeze structure during the transition periods at the southern Bulgarian Black sea coast. Atmos Environ 37: 299–311

Smalikho IN, Köpp F, Rahm S (2005) Measurement of atmospheric turbulence by 2-μm Doppler lidar. J Atmos Oceanic Technol 22:1733–1746

Spänkuch D, Döhler W, Güldner J, Keens A (1996) Ground-based passive atmospheric sounding by FTIR emission spectroscopy. First results with EISAR. Beitr Phys Atmos 69: 97–111

Steyn DG, Baldi M, Hoff RM (1999) The detection of mixed layer depth and entrainment zone thickness from lidar backscatter profiles. J Atmos Ocean Technol 16:953–959

Stull RB (1988) An introduction to boundary layer meteorology. Kluwer, Dordrecht

Stutz J, Alicke B, Ackermann R, Geyer A, White A, Williams E (2004) Vertical profiles of NO_3, N_2O_5, O_3, and NO_x in the nocturnal boundary layer: 1. Observations during the Texas Air Quality Study 2000. J Geophys Res 109: D12306
Sun J, Burns SP, Lenschow DH, Banta R, Newsom R, Coulter R, Frasier S, Ince T, Nappo C, Cuxart J, Blumen W, Lee X, Hu X-Z (2002) Intermittent turbulence associated with a density current passage in the stable boundary layer. Bound-Lay Meteor 105:199–219
Sun J, Lenschow D, Burns S, Banta R, Newsom R, Coulter R, Frasier S, Ince T, Nappo C, Balsley B, Jensen M, Mahrt L, Miller D, Skelly B (2004) Intermittent turbulence in stable boundary layers and the processes that generate it. Bound-Lay Meteor 110:255–279
Troen I, Petersen EL (1989) European Wind Atlas. Risø National Laboratory for the commission of the European communities, Roskilde, ISBN 87-550-1482-8, p. 656
Uno I, Wakamatsu S, Ueda H, Nakamura A (1988) An observational study of the structure of the nocturnal urban boundary layer. Bound-Lay Meteor 45:59–82
Uno I, Wakamatsu S, Ueda H, Nakamura A (1992) Observed structure of the nocturnal urban boundary layer and its evolution into a convective mixed layer. Atmos Environ 26B: 45–57
Vaughan G, Wareing DP, Pepler SJ, Thomas L, Mitev VM (1993) Atmospheric temperature measurements made by rotational Raman scattering. Appl Optics 32:2758–2764
Vogt S, Engelbart D (2002) Heat flux measurements with a boundary-layer profiler and comparison with other instruments. Proceedings of the 11th ISARS, Rome;343–345
Vogt S, Jaubert G (2004) Föhn in the Rhine valley as seen by a wind-profiler-RASS System and comparison with the non-hydrostatic model Meso-NH. Meteorol Z 13:165–174
Weissmann MD, Mayr GJ, Banta RM, Gohm A (2004) Observations of the temporal evolution and spatial structure of the gap flow in the Wipp valley on 2 and 3 October 1999. Mon Wea Rev 132:2684–2697
Wesely M (1976) The combined effect of temperature and humidity on the refractive index. J Appl Meteorol 15:43–49
Westwater ER, Han Y, Irisov VG, Leuskiy V, Kadygrov EN, Viazankin SA (1999) Remote sensing of boundary layer temperature profiles by a scanning 5-mm microwave radiometer and RASS: Comparison experiments. J Atmos Oceanic Technol 16:805–818
White AB, Senff CJ, Banta RM (1999) A comparison of mixing depths observed by ground-based wind profilers and an airborne lidar. J Atmos Oceanic Technol 16:584–590
Wilczak JM, Gossard EE, Neff WD, Eberhard WL (1996) Ground-based remote sensing of the atmospheric boundary layer: 25 years of progress. Bound-Lay Meteorol 78:321–349
Wulfmeyer V (1999) Investigation of turbulent processes in the lower troposphere with water-vapor DIAL and Radar-RASS. J Atmos Sci 56:1055–1076
Wulfmeyer V, Janjić T (2005) 24-h observations of the marine boundary layer using ship-borne NOAA high-resolution Doppler lidar. J Appl Meteorol 44:1723–1744
Wulfmeyer VO, Randall M, Brewer WA, Hardesty RM (2000) 2 μm Doppler lidar transmitter with high frequency stability and low chirp, Opt Lett 25:1228–1230

Chapter 5
Outlook

5.1 Introduction

Remote sensing, either ground-based or airborne or space-borne, is a still developing field in instrument development and atmospheric boundary layer research. Thus, the preceding chapters have been merely a snapshot of the present state. The ongoing enhancements are mainly driven from two sides: technological advancements and application needs.

5.2 Technological Advancements

Advancements in electronics and microelectronics, in laser techniques, and in computer resources allow for increasingly complex sounding, detection, and evaluation methods. Progress is especially made in the field of optical and electromagnetic (including microwave) remote sensing (see e.g. Gaffard et al. 2008) and ever-growing data evaluation procedures (Rodgers 2000).

For instance, the sensing of optical polarization information in addition to the sensing of radiation intensity in remote sensing is increasingly used in atmospheric sounding. Imaging polarimetry is an emerging technique that promises to enhance many fields of optical metrology ranging from remote sensing to atmospheric sciences to industrial monitoring. Polarimetry means to measure information about the vector nature of the optical field across a scene. Imaging polarimetry is a special case of polarimetry that is dedicated to mapping the state of polarization across a scene of interest. A review of the potentials of this technique is given in Tyo et al. (2006). Polarimetry is used, e.g. to determine the amount, size, and type of aerosols.

Laser technology also offers better measurement options today. For example, recently a coherent Doppler LIDAR at 2 μm wavelength has been built with higher output energy (100 mJ) than previously available. The laser transmitter is based on diode-pumped Ho:Tm:LuLiF, a newly developed laser material that allows more efficient energy extraction. Single-frequency operation is achieved

S. Emeis, *Surface-Based Remote Sensing of the Atmospheric Boundary Layer*, Atmospheric and Oceanographic Sciences Library 40, DOI 10.1007/978-90-481-9340-0_5,

by a ramp-and-fire injection seeding technique. An advanced photodetector architecture is used incorporating photodiodes in a dual-balanced configuration. This Doppler LIDAR allows for real-time display of wind and aerosol backscatter data products. The high-pulse energy and receiver efficiency provides for measurement of wind fields to ranges not seen before with 2 μm LIDARs (Koch et al. 2007).

Evaluation and comparison of new sounding instruments was performed, e.g. in the international COST 720 Temperature, hUmidity, and Cloud (TUC) profiling experiment organized in Winter 2003/2004 at Payerne, Switzerland. Various in situ and active/passive ground-based remote-sensing systems, including three microwave radiometers, a cloud RADAR, and a windprofiler, were operated in the same location. Temperature and humidity profiles from two microwave radiometers were compared with radiosonde profiles (Klaus et al. 2006).

Still increasing computer resources allow for a more rapid processing of the incoming data, e.g. with spectrometers and interferometers. Larger storing devices permit the prolongation of maintenance intervals. Less energy consumption of modern electronics cordless data transfer options make more remote installations of the instruments possible.

5.3 Application Needs

Several new challenges and demands from the application side have been brought up to the scientific community dealing with boundary layer remote sensing in the past few years. The latter demands comprise among other issues (i) meteorological input for planning and conducting electricity generation from renewable natural sources (wind, solar, water power, Emeis et al. 2007; Frehlich and Kelley 2008), (ii) the interrelation between atmospheric boundary layer meteorology and the emission of greenhouse gases (see Laj et al. 2009 for an update of all kinds of atmospheric composition measurement techniques including ground-based remote sensing), and (iii) the interrelation between boundary layer air quality and public health. Here, especially the aerosol load of near-surface air has got into the focus. See for example Wu et al. (2007) for the use of FTIR spectroscopy for aerosol detection or Schäfer et al. (2009) for the role of remotely detected mixing layer heights for the assessment of aerosol concentrations.

The first application demand requires, e.g., very accurate wind measurements for the operation of wind energy converters, accurate shortwave radiation measurements for the operation of photovoltaic solar energy converters, and accurate areal precipitation measurements for the exploitation of water power. These accurate data are not only needed for an estimation of the gross amount of the expected energy output but rather as input values for a short-term prediction of the temporal variation of the energy output. With the growing fraction of renewable energies in electricity generation, the perfect organization of power grids connecting producers and consumers becomes increasingly important, because the production and the consumption of electrical energy must be balanced at any time.

A special challenge in boundary layer remote sensing connected to the generation of electricity from the wind is the exploration of the wakes of single wind turbines as well as of entire wind parks. The analysis of these wakes is essential for the assessment of the influence of a windward wind turbine on further turbines downwind in a wind park. The result of such assessments is an optimized distance of turbines in a wind park. Likewise, wind parks affect each other if they are too close together. Therefore, not only the turbine distance within a park must be tuned but also the distance between entire wind parks must be optimized for most effective use of the available space for wind energy conversion (Emeis 2010). Main features of such wakes are reduced wind speeds and enhanced turbulence intensities in limited areas behind the disturbing objects. Their analysis requires spatially and temporally highly resolved wind profile measurements. Thus, optical techniques are probably the most ideal methods for this task. First experiments have been made in recent years with small wind LIDARs, but much further work on this issue is necessary. One example is the operation of a horizontally looking scanning wind LIDAR mounted on the nacelle of a wind turbine that scans the wake of the turbine (Bingöl et al. 2007).

First attempts have been performed to obtain the three-dimensional structure of the wakes by simultaneous observation with two or even three ground-based wind LIDARs. For that purpose, three pulsed LIDARs were used in staring mode, placed so that their beams cross close to a 3D sonic anemometer mounted at a mast 78 m above the ground. The results show generally very good correlation between the LIDAR and the sonic times series, except that the variance of the velocity measured by the LIDAR is attenuated due to spatial filtering. The amount of attenuation can, however, be predicted theoretically by use of a spectral tensor model of the atmospheric surface layer turbulence (Mann et al. 2009).

A further, tightly related example is the optimal operation of a wind turbine in a turbulent wind field. Optimizing turbine orientation and blade angles requires short-term prediction on upcoming turbulence elements in the incident flow. Suggestions have been made to operate small wind LIDARs on turbine nacelles, which look horizontally into the approaching wind flow (e.g. Harris et al. 2007).

The second application demand that fosters further instrument development is monitoring of greenhouse gas emissions. With progressing climate change, the supervision of these emissions increasingly becomes a major task. Vertical profiling of the diurnal variation of such gas concentrations could provide input for determining the source strengths of these gases via budget methods (Denmead et al. 1996, 2000; Emeis 2008). Direct emission measurements will be increasingly made from turbulent flux determination.

The third application demand requires among other tasks concentration measurements and the determination of ABL features, which limit the dispersion of trace gases. Two of these ABL features are the mixing-layer height and the vertical wind profile above the surface layer. The variables "mixing layer height" and "atmospheric boundary layer height" are important for the description of vertical profiles within the boundary layer (Peña et al. 2008, 2010) and for air quality assessments (e.g. Emeis et al. 2008). These variables can neither be measured nor modelled directly. The present state of detecting these heights has been described in Chapter 4, but all the listed methods still have considerable limitations. So efforts are ongoing to develop reliable monitoring strategies for these heights based on ground-based remote sensing.

But also nowcasting, the weather forecast for the next hour, and monitoring of near-surface weather features which have immediate effects on issues such as air traffic security on airports (wind gusts, downdraughts, cross-winds, fog, snow) lead to an ongoing development of ground-based remote-sensing techniques. Recently, first LIDARs for wake vortex detection have been mounted on aircraft. On-board axial detection of wake vortices is more complex than the classic ground-based tangential approach, because the axial air speed in these vortices is low and the atmospheric particle density is reduced, yielding a poorer signal-to-noise ratio (Douxchamps et al. 2008).

Determination of the height of cloud and fog tops is a subject of current research as well. In the abovementioned COST 720 TUC experiment, this problem has been addressed, too. While the base layer of clouds is well measured with ceilometers, the top of the cloud was determined in this experiment using a 78 GHz FMCW radar. This method for cloud top determination is based on determining the height at which the radar signal falls below a defined threshold level and is still under development (Ruffieux et al. 2006).

Additionally, new monitoring techniques, which deliver nearly continuous information on profiles of atmospheric variables over larger parts of the complete ABL, in turn trigger the further elaboration of theoretical concepts and analytical and numerical description of various features of the ABL. Recent updates on techniques and methods for ground-based remote sensing are presented and discussed on meetings and workshops. Especially devoted to this special subject is the biennial series of the "International Symposium for the Advancement of Boundary Layer Remote Sensing (ISARS)" (selected papers from the 2006 and 2008 ISARS can be found in Meteorologische Zeitschrift (Emeis 2007; Mann and Emeis 2009)). See Appendix for more details on this conference series. But also other scientific events such as the triennial series of the "International Symposium on Tropospheric Profiling (ISTP)" deal with new developments in this field. The full proceedings from the last conference of this latter series in 2009 are available online.

References

Bingöl F, Larsen GC, Mann J (2007) Wake meandering – An analysis of instantaneous 2d laser measurements. J Phys Conf Ser 75:012059. doi: 10.1088/1742-6596/75/1/012059

Denmead OT, Raupach MR, Dunin FX, Cleugh HA, Leuning R (1996) Boundary-layer budgets for regional estimates of scalar fluxes. Global Change Biol 2:255–264

Denmead OT, Leuning R, Griffith DWT, Jamie IM, Esler MB, Harper LA, Freney JR (2000) Verifying inventory predictions of animal methane emissions with meteorological measurements. Bound-Layer Meteor 96:187–209

Douxchamps D, Lugan S, Verschueren Y, Mutuel L, Macq B, Chihara K (2008) On-board axial detection of wake vortices using a 2-μm LiDAR. IEEE Trans Aerospace Electronic Sys 44:1276–1290

Emeis S (2007) ISARS 13 special issues. Meteorol Z 16:323–324
Emeis S (2008) Examples for the determination of turbulent (sub-synoptic) fluxes with inverse methods. Meteorol Z 17:3–11
Emeis S (2010) A simple analytical wind park model considering atmospheric stability. Wind Energy 13:459–469
Emeis S, Harris M, Banta RM (2007) Boundary-layer anemometry by optical remote sensing for wind energy applications. Meteorol Z 16:337–347
Emeis S, Schäfer K, Münkel C (2008) Surface-based remote sensing of the mixing-layer height – a review. Meteorol Z 17:621–630
Frehlich R, Kelley N (2008) Scanning Doppler lidar measurements for wind energy applications. Eos Trans AGU 89(53): Fall Meet Suppl Abstract A21E–0225
Gaffard C, Nash J, Walker E, Hewison TJ, Jones J, Norton EG (2008) High time resolution boundary layer description using combined remote sensing instruments. Ann Geophys 26:2597–2612
Harris M, Bryce DJ, Coffey AS, Smith DA, Birkemeyer J, Knopf U (2007) Advance measurement of gusts by laser anemometry. J Wind Eng Industr Aerodyn 95:1637–1647
Klaus V, Bianco L, Gaffard C, Matabuena M, Hewison TJ (2006) Combining UHF radar wind profiler and microwave radiometer for the estimation of atmospheric humidity profiles. Meteorol Z 15:87–98
Koch GJ, Beyon JY, Barnes BW (2007) High-energy 2 μm Doppler lidar for wind measurements. Opt Eng 46:116201. doi: 10.1117/1.2802584
Laj P, Klausen J, Bilde M, Plaß-Duelmer C, Pappalardo G, Clerbaux C, Baltensperger U, Hjorth J, Simpson D, Reimann S, Coheur P-F, Richter A, De Mazière M, Rudich Y, McFiggans G, Torseth K, Wiedensohler A, Morin S, Schulz M, Allan JD, Attié J-L, Barnes I, Birmili W, Cammas JP, Dommen J, Dorn H-P, Fowler D, Fuzzi S, Glasius M, Granier C, Hermann M, Isaksen ISA, Kinne S, Koren I, Madonna F, Maione M, Massling A, Moehler O, Mona L, Monks PS, Müller D, Müller T, Orphal J, Peuch V-H, Stratmann F, Tanré D, Tyndall G, Abo Riziq A, Van Roozendael M, Villani P, Wehner B, Wex H, Zardini AA (2009) Measuring atmospheric composition change. Atmos Environ 43:5351–5414
Mann J, Emeis S (2009) ISARS 2008 special issues. Meteorol Z 18:123–124
Mann J, Cariou J-P, Courtney MS, Parmentier R, Mikkelsen T, Wagner R, Lindelöw P, Sjöholm M, Enevoldsen K (2009) Comparison of 3D turbulence measurements using three staring wind lidars and a sonic anemometer. Meteorol Z 18:135–140
Peña A, Gryning S-E, Hasager CB (2008) Measurements and Modelling of the Wind Speed Profile in the Marine Atmospheric Boundary Layer. Bound-Lay Meteor 129:479–495
Peña A, Gryning S-E, Hasager CB (2010) Comparing mixing-length models of the diabatic wind profile over homogeneous terrain. Theor Appl Climatol online first. doi: 10.1007/s00704-009-0196-8
Rodgers CD (2000) Inverse Methods for Atmospheric Sounding: Theory and Praxis. World Scientific Publishing Co, Singapore, ISBN 981-02-2740-X
Ruffieux D, Nash J, Jeannet P, Agnew JL (2006) The COST 720 temperature, humidity, and cloud profiling campaign: TUC. Meteorol Z 15:5–10
Schäfer K, Wang Y, Münkel C, Emeis S, Xin J, Tang G, Norra S, Schleicher N, Vogt J, Suppan P (2009) Evaluation of continuous ceilometer-based mixing layer heights and correlations with PM2.5 concentrations in Beijing. In: Picard RH, Schäfer K, Comeron A, Kassianov EI, Mertens CJ (Eds). Remote Sensing of Clouds and the Atmosphere XIV. Proceedings of SPIE 7475: 74750 N-74750 N-12, http://spie.org
Tyo JS, Goldstein DL, Chenault DB, Shaw JA (2006) Review of passive imaging polarimetry for remote sensing applications. Appl Opt 45:5453–5469
Wu, C-F, Chen, Y-L, Chen, C-C, Yang, T-T, Chang, P-E (2007) Applying open-path Fourier transform infrared spectroscopy for measuring aerosols, J Environ Sci Health Part A 42:1131–1140

Appendix

A.1 Introduction

ISARS, the "International Society of Acoustic Remote Sensing of the Atmosphere and Oceans", is an informal association of scientists from all over the world. Its main activity is the organization of the series of conferences, which were originally called "International Symposia on Acoustic Remote Sensing and Associated Techniques of the Atmosphere and Oceans" (also abbreviated ISARS). The Society was founded in 1979 in Boulder, and the first ISARS conference was held in Calgary in 1981. The society meets and holds elections at every ISARS conference. The idea behind this conference series is to provide a regular forum for researchers to present recent developments in remote-sensing technology and applications to the atmospheric boundary layer. Furthermore, the conferences have always been an opportunity to bring manufactures and scientists together. The intention is that scientists from all over the world should participate and the cost should be held relatively low.

A.2 Name and Scope

While the abbreviation ISARS has remained unchanged for the society and for its conferences since Calgary 1981, the words for the name of the conferences have been changed from "International Symposium on Acoustic Remote Sensing and Associated Techniques of the Atmosphere and Oceans" to "International Symposium for the Advancement of Boundary Layer Remote Sensing (ISARS)". This decision was taken at the 12th ISARS conference in Cambridge in 2004, because other techniques than the acoustic ones have also become important for boundary layer remote sensing in the preceding years. Therefore, it had been decided in Cambridge to enlarge the scope of ISARS and to include also optical and electromagnetic techniques for boundary layer remote sensing. The scope of the ISARS conferences now comprises SODARs, LIDARs, microwave radiometers, and boundary layer wind profilers. This broadening widens the spectrum of applications discussed at ISARS conferences. In former times, when concentrating

S. Emeis, *Surface-Based Remote Sensing of the Atmospheric Boundary Layer*,
Atmospheric and Oceanographic Sciences Library 40, DOI 10.1007/978-90-481-9340-0,

on wind and turbulence profiles and qualitative aspects of the atmospheric layering, dynamic aspects of boundary layer meteorology had been in the focus of these meetings. This field is still important, and with the growing demand for wind energy as a renewable energy source, it has found a new challenging subject for further developments. Specifically, lasers for remote wind sensing are developing rapidly. But with the steadily increasing possibility to measure temperature and aerosol profiles with electromagnetic and optical remote-sensing techniques, thermodynamic and air pollution aspects are growingly added to the agenda of the ISARS conferences.

A.3 The Series of Conferences

After the inaugurational meeting in Calgary, the symposium has been held in Rome (1983), Paris (1985), Canberra (1988), New Delhi (1990), Athens (1992), Boulder (1994), Moscow (1996), Vienna (1998), Auckland (2000), Rome (2002), Cambridge (2004), Garmisch-Partenkirchen (2006, the first time with the new name), and in Risø near Roskilde (2008). The latest conference took place in Paris at the end of June 2010 and the next one is scheduled for 2012. Thus, this conference series now runs successfully for nearly 30 years.

A review on the first ten conferences has been provided by Singal (2000). The 11th conference in 2002 was hosted by the Institute of Atmospheric Sciences and Climate (ISAC) of the Italian National Research Council (CNR) and took place in Rome, Italy, on June 24–28, 2002. The organisation was run by Stephania Argentini and Giangiuseppe Mastrantonio. In 2004, Philip Anderson from the British Antarctic Survey and Sabine von Hünerbein and Stuart Bradley from Salford University prepared and performed the 12th ISARS conference in the perfect old and dignified surroundings of Claire College in Cambridge, UK.

The 13th ISARS conference in 2006 was organised by Stefan Emeis from the Institute of Meteorology and Climate Research – Department on Atmospheric Environmental Research – of the then Forschungszentrum Karlsruhe, situated in Garmisch-Partenkirchen. This Forschungszentrum Karlsruhe is now part of the Karlsruhe Institute of Technology since October 2009. About 65 participants from 14 countries came to the 13th meeting. In four different sessions, 46 oral and about 25 poster presentations were given. The four sessions dealt with the theory of atmospheric and oceanic boundary layer remote sensing, with new remote-sensing techniques apart from the sodar technology (i.e. optical and electro-magnetic remote-sensing techniques), with applications of boundary layer remote sensing, and with the comparison of remote-sensing information with model results. The session on applications included a special part devoted to wind energy issues. The decision taken in Cambridge 2 years before proved to be a success. Nearly one quarter of the presentations dealt with optical and electromagnetic remote sensing. So, very modern developments of boundary layer remote-sensing techniques were discussed in this meeting. The number of participants was about 15% higher compared to the preceding ISARS symposium. The conference was partly sponsored by the German Meteorological Society and by the company METEK. The European Meteorological Society supported the meeting, and a Young Scientist Travel Award was given to a young Russian participant, Rostislav Kouznetsov.

In 2008, at the 14th conference in Risø near Roskilde, organised by Jacob Mann from Risø National Laboratory (a part of the Danish Technical University, DTU), participants from every inhabitable continent of the Earth except Africa took part: 18 countries were represented by 82 participants, which is an increase of 25% relative to the 13th ISARS. The participants provided 39 oral presentations and 40 posters, which were accompanied by 67 scientific papers published by Institute of Physics as the first volume of IOP Conference Series: Earth and Environmental Science (permanently available from: www.iop.org/EJ/toc/1755-1315/1/1).

The last, 15th conference took place in Paris in 2010, a place where it had been already 25 years ago. It was organised by Martial Haeffelin from the Site Instrumental de Recherche par Télédétection Atmosphérique (SIRTA) at the Institut Pierre Simon Laplace (IPSL) in Palaiseau near Paris. The number of participants roughly doubled compared to the 14th conference and more than 120 presentations were delivered. This shows that this series of conferences has found its market.

Table A.1. Overview of ISARS conferences

Place	Year	Publications (no entry: on-site conference proceedings only)
Calgary, Canada	1981	
Rome, Italy	1983	
Paris, France	1985	*Atmospheric Research*, Vol. 20 (1986), Nos. 2–4, pp. 105–333
Canberra, Australia	1988	
New Delhi, India	1990	Singal SP (ed.) *Acoustic Remote Sensing. Proceedings of the 5th Int. Symp. of Acoustic Remote Sensing of the Atmosphere and Oceans*. Tata McGraw Hill Publ. Comp. Ltd., New Delhi, 538 pp.
Athens, Greece	1992	*Applied Physics*, Vol. B57 (1993) No. 1, pp. 1–82. *International Journal of Remote Sensing*, Vol. 15 (1994) No. 2.
Boulder, USA	1994	*Boundary-Layer Meteorology*, Vol. 81 (1996) No. 1, pp. 1–103.
Moscow, USSR	1996	
Vienna, Austria	1998	*Meteorology and Atmospheric Physics*, Vol. 71 (1999) No. 1–2, pp. 1–138
Auckland, New Zealand	2000	
Rome, Italy	2002	*Meteorology and Atmospheric Physics*, Vol. 85 (2004), No. 1–2, pp. 1–186
Cambridge, UK	2004	
Garmisch, Germany	2006	*Wissenschaftliche Berichte FZKA*, Vol. 7222 (2006), 186 pp *Meteorologische Zeitschrift*, Vol. 16 (1997), No. 4, pp. 323–459 *Meteorologische Zeitschrift*, Vol. 16 (1997), No. 5, pp. 467–523
Risø, Denmark	2008	*IOP Conference Series: Earth and Environmental Science*, Vol. 1 (2008), www.iop.org/EJ/toc/1755-1315/1/1 *Meteorologische Zeitschrift*, Vol. 18 (1999), No. 2, pp. 123–195 *Meteorologische Zeitschrift*, Vol. 18 (1999), No. 3, pp. 235–337
Paris, France	2010	planned to appear in *Boundary-Layer Meteorology*

Reference

Singal SP (2000) Reflections on ISARS: a history of 10 ISARS symposia. Proceedings of the 10th ISARS 2000, Auckland, pp. 1–6

Glossary

Absorption Uptake of energy without backscatter or fluorescence. The gained energy is radiated away as thermal radiation later. Wavelength-dependent absorption of radiation by trace gases may be utilized to measure the concentration of trace gases in the atmosphere.

Active remote sensing Remote-sensing methods based on the emission of a discontinuous (pulsed) or continuous signal and the subsequent receiving of the backscattered signal are called active remote-sensing methods. The opposite method is passive remote sensing.

Acoustic temperature The speed of sound c depends on air pressure p, air density ρ and the ratio $\gamma = c_p/c_v$, i.e. $c = \sqrt{\gamma \frac{p}{\rho}} = \sqrt{\gamma RT} \approx 20.063\sqrt{T}$ with the specific gas constant R, the specific heat for constant pressure, c_p and the specific heat for constant volume, c_v. As γ and R slightly depend on atmospheric humidity, the speed of sound is humidity-dependent as well. The temperature, T_a, which, when entered into the definition for the speed of sound for dry air, gives the true sound speed in humid air, is called acoustic temperature. This temperature can be computed from the dry-bulb temperature T and the specific humidity q by $T_a = T + 0.51q$ (Schotanus et al. 1983). It can be determined directly from measurements of the sound speed. Its definition is similar to the one for the virtual temperature. In cold and dry environments, the differences between acoustic and virtual temperature are negligible, in warm and moist environments the differences become important.

Anti-Stokes scattering see "Raman scattering".

Aperture Aperture usually defines the diameter of the opening of a optical system.

Attenuation Attenuation is the gradual depletion of energy along the propagation path of a signal, usually due to absorption or scattering. It is sometimes also called extinction.

Backscatter Scattering back into the direction from which the signal has arrived is called backscatter.

Baffles Lateral shields which protect the environment from acoustic signals coming from a SODAR are called baffles.

Beam focusing A method to determine the distance between an active optical remote-sensing instrument and the backscattering air volume. The optical beam is focused in some distance from the instrument. The strong backscatter from the range around the focus is dominating the received signal. Thus the retrieved information is attributed to this focal range. The focal distance can be varied within certain bounds. This technique allows for continuous (non-pulsed) observations. Alternatives are the measurement of travel times for pulsed active remote-sensing techniques and techniques using frequency modulation for continuous wave remote sensing.

Bistatic A remote-sensing method, which is based on a spatial separation of emitter and receiver is called a bistatic method. This often requires power supply in both places. The opposite method is called monostatic.

Bragg condition Optimal backscatter is received, if the scattering objects in the atmosphere have about half the size of the wavelength of the emitted optical, electromagnetic, or acoustic radiation. The Bragg condition that is essential for optimal atmospheric remote sensing has been described for the first time in 1913 by the British physicists William Henry Bragg (1862–1942) and William Lawrence Bragg (1890–1971) shortly after their return from Adelaide, South Australia, where W.H. Bragg was a professor at Adelaide University from 1885 until 1908, by referring to the scattering of x-rays at crystals (Bragg and Bragg 1913). Backscatters from different centres of action have a maximum positive interference, if this condition is met.

Brightness temperature The brightness temperature is the temperature of a black body (emissivity equals unity), which is in thermal equilibrium with its surroundings. For grey bodies (emissivity less than unity) the brightness temperature is lower than the true temperature of the body.

Cabannes line The resulting intensity distribution of elastically scattered radiation (see "Rayleigh scattering") is called Cabannes line. It has a spectral width of a few GHz or about 0.1 cm^{-1}. If detection is made with low spectral resolution, the sum of the scattering at the Cabannes line and the Raman lines (see "Raman scattering") is called Rayleigh scattering. From the point of view of Raman scattering, the Cabannes line is the central line between the lines from Stokes and Anti-Stokes scattering (She 2001, Wandinger 2005). The scattering process leading to this sharp central line is called Cabannes scattering in contrast to the more general term Rayleigh scattering.

Chaff Chaff is artificial reflecting material dispersed in the air from aircraft in order to enhance or (for military purposes) to disturb electromagnetic backscatter from the atmosphere. Chaff can be, e.g. made from thin pieces of aluminium or metalized glass fibres or plastic. Due to its low sedimentation rate, it may be used as atmospheric tracer.

Coherent Coherent or heterodyne determination means that the received backscattered signal is mixed with a signal from a second (local) oscillator before analysis in the receiving instrument. The resultant beat frequency is much lower than the frequency of the backscattered signal and can easily be analysed. This is one of two principal techniques used to analyse the Doppler shift of backscattered light. See also “incoherent”.

Continuous-wave Often RADARs are operated as continuous-wave RADARs (CW-RADAR) as an alternative to pulsed RADARs. CW-RADARs continuously emit signals. The advantage of CW-RADARs is that they need considerably less power and shielding efforts. The disadvantage of CW-RADARs is that they do not allow for an easy range determination via the delay between emitted and received signal.“frequency-modulated continuous-wave”.

Differential absorption Trace gas-specific active optical remote sensing can be performed by emitting two signals with neighbouring wavelengths. One wavelength is chosen so that it coincides with the centre of an absorption line of this trace gas, the other close-by wavelength is chosen besides this absorption line. The difference of the backscattered signal depends only on the trace gas concentration, because backscatter from aerosols is practically identical for both wavelengths.

Doppler beam swinging method The Doppler beam swinging (DBS) method is used to measure all three spatial components of the wind vector. Here, three to five pulses are emitted one after the other into three to five different directions and the Doppler shift of the backscattered signal is analyzed. One of the three or five directions is usually the vertical direction, the other directions are tilted by 15° to 20° from the vertical. The azimuthal directions of the tilted beams usually differ by 90°. The angle between the slanted beams should be as low as possible to receive the three – or five – different pieces of information, which are later used to compute the three orthogonal wind components, from air volumes close together. If the angle becomes too small, the computation of horizontal wind components from nearly vertical radial velocity components along the beams is too uncertain.

Because the next pulse cannot be emitted before the backscatter of the previously emitted pulse has been received in order to avoid disturbances, one measurement cycle lasts six to ten times the time the pulse needs to travel up to the maximum vertical range of the instrument. A related technique is conical scanning (see “scanning”).

Doppler broadening Caused by the thermal motion of gas molecules spectral lines are not ideally thin but have a certain width due to the many small Doppler shifts (“Doppler effect”) from the moving molecules. The shape of a Doppler broadened line is Gaussian. The intensity of the thermal motion depends on the temperature of the gas. Thus, Doppler broadening is temperature-dependent and the determination of the line width may be used for temperature measurements. This requires spectrally high resolved measurements.

Doppler effect The Doppler frequency shift Δf (named after the Austrian physicist C. Doppler (1803–1853)) is proportional to the emitted frequency f_0 and to the radial speed of the scattering object, v where $\Delta f \approx 2 f_0\, v / c$. The relative frequency shift, $\Delta f/f_0$, depends only on the ratio between the velocity of the scattering object and the propagation speed, c, of the emitted signal. The received frequency is increased (the wavelength is reduced) if the scattering air volume moves towards the instrument and it is reduced if the air moves away.

Dual-Doppler technique Making wind speed measurements with two Doppler RADARs or Doppler LIDARs within the same air volume from two different positions is called Dual-Doppler technique. While a single instrument delivers the radial wind component along the line-of-sight only, two instruments allow for the determination of both horizontal wind components simultaneously without involving areal averages.

Echogram Time-height cross-sections of the acoustic backscatter intensity measured from SODAR are often called echograms or facsimile plots.

Elastic scattering Elastic scattering is Mie or Rayleigh scattering. Apart from the Doppler shift, the wavelength of the backscattered signal is equal to the emitted wavelength. See also "inelastic scattering".

Facsimile plot see "echogram".

Fixed echoes Often, when the emitted beam is not well focused (side lobes are present), backscatter from solid objects away from the path of the emitted beam can disturb the analysis of the received signal. The term has been coined for acoustic remote sensing. Fixed echoes are seen as enhanced backscatter intensities in certain range intervals, which correspond to the distance of the object causing the fixed echoes. If wind speeds are analyzed from acoustic backscatter, fixed echoes lead to too low wind speeds. In RADAR meteorology the term "ground clutter" is used frequently.

Fluorescence Fluorescence happens when incoming radiation of a certain frequency excites an atom or molecule and brings it into a resonance state. When the atom or molecule returns to its ground state, light of another specific frequency is emitted. As fluorescence is obstructed by collisions of the air molecules among each other, fluorescence methods can only be employed for the detection of trace substances in the upper atmosphere above about 15 km, but not in the boundary layer. Fluorescence is often much more intense than Raman scattering.

Frequency-modulated continuous-wave Continuous-wave RADARs cannot determine the range from the travel time, i.e. the time delay between emitted and received signal. Beam focusing is not applicable either. Therefore, frequency-modulated continuous-wave RADARs (FMCW-RADAR) emit a signal with constantly changing frequency. The range of the scattering object in the line of sight can then be determined from the frequency difference between the emitted and the received signal.

Groundclutter see "fixed echoes".

Heterodyne see "coherent".

Imaging Imaging is the process of measuring so as to build an extensive two-dimensional description of the received properties; comparable to a classical photograph. This is often done by scanning in azimuth and elevation or by a system of optical or electromagnetic lenses.

Incoherent Direct or incoherent determination means that the frequency of received backscattered light is analyzed directly with a high-resolution spectrometer or with a interferometer. This is one of two principal techniques used to analyse the Doppler shift of backscattered light. It is constructionally simple, but it requires the use of narrow-band optical filters through which the radiation must pass. See also "coherent".

Inelastic scattering Raman scattering is inelastic scattering. The wavelength of the backscattered signal is different from the wavelength of the incoming signal. See also "elastic scattering".

In-situ measurements Measurements which record the state of the analyzed object in the direct vicinity of the instrument are called in situ measurements. Usually, a direct contact between the measured object and the sensor of the instruments occurs. In situ measurements may be influenced by a change in the state of the measured object due to the presence of the instrument. The opposite method is remote sensing.

Interferometer An interferometer records an interferogram by mixing two beams which had different travel distances. It often needs laborious techniques such as Fast Fourier Transforms to derive ordinary wavelength or frequency spectra from interferograms. The advantage is the much higher signal intensity of the received information. Grates or prisms in normal spectrometers allow only for much less signal intensities.

Mie scattering Mie scattering is a form of elastic scattering that occurs if the size of the scattering particle (molecule, aerosol, droplet, insect) has about the same wave length as the scattered radiation. Ground-based active remote sensing is usually relying on Mie scattering, because obeying the Bragg condition means to have comparable dimensions of the wavelength of the emitted radiation and the scattering object. Light scattering in mist or fog is usually Mie scattering.

Monostatic A remote-sensing method, which is based on having emitter and receiver in the same place is called a monostatic method. This method is simpler than the bistatic method, because power supply is required in one place only. The backscattering of the emitted signal to the receiver may either be performed by natural scattering processes or by special mirrors mounted in some distance. The opposite method is called bistatic.

Passive remote sensing Remote-sensing methods based on the receiving of naturally emitted radiation are called passive remote-sensing methods. The opposite method is active remote sensing.

Path-averaging Path-averaging remote sensing delivers a mean signal, which is representative for the whole path length over which the signal has travelled. No range resolution is possible. Usually, a continuous signal is analyzed. There are active and passive path-averaging methods available. The opposite technique is profiling or sounding. See also "tomography".

Phased-array antenna Phased-array SODARs have a larger number of smaller sound transducers (often 32 or 64), which are regularly arranged on a quadratic plate with an area of about 1 m^2. By using interference principles (enhancement and extinction), slanted and vertical beams are formed by operating the various sound transducers with specified time delays among each other. While with classical SODARs the antennas simultaneously serve as a shield to protect the environment from the sound pulses as well as the SODAR from disturbing ambient noise, phased-array SODARs need a shield around them.

Polarization Polarization means the orientation of the plane within which a wave oscillates. This plane is perpendicular to the propagation direction of the wave. Certain scattering processes lead to well-defined changes in the orientation of the scattered waves, which can be used to identify properties of the scatterer.

Pressure broadening Collision between emitting particles or molecules disturbs the emission process. This leads to non-zero line widths. The number of collisions increases with increasing pressure. The shape of pressure broadened emission lines is Lorentzian.

Profiling see "sounding".

Pulsed RADAR RADARs which emit discontinuous signals (pulses of given length and frequency) are called pulsed RADARs in contrast to continuous-wave RADARs. Pulsed RADARs can easily determine the range from the time delay between the emitted and the received signal.

Radome A radome (from the words RADAR and dome) is a usually spherical weatherproof enclosure which protects RADAR antennas.

Raman scattering With normal elastic scattering, a molecule absorbs a photon, enters into an excited vibrational or rotational state, and immediately emits (backscatters) the photon again when returning to its ground state. With inelastic Raman scattering, the molecule does not return to the ground state, but to a neighbouring energy level so that a photon with a slightly changed (usually greater) wavelength is emitted (Stokes scattering). If the emitted wavelength is lower, the process is called Anti-Stokes scattering. Raman scattering is distinguished from fluorescence by that it can happen for radiation of any wavelength and the excited molecule does not enter into a resonance state.

The observed wavelength shift due to the excitation of rotation or vibration of air molecules is characteristic for a given trace gas species. This scattering phenomenon is named after the Indian physicist Chandrasekhara Venkata Raman, who first reported the experimental discovery of this wavelength shift in 1928. The disadvantage of the method is that inelastically backscattered signal intensities are two to three orders of magnitude lower than those from elastic Rayleigh scattering.

Range gate Height interval to which a received signal will be attributed. The size of the range gate depends on the signal propagation speed and pulse length (pulsed measurements) or on the frequency resolution (continuous-wave measurements).

Rayleigh scattering Rayleigh scattering is a form of elastic scattering that occurs if the size of the scattering object is much smaller than the wave length of the scattered radiation. The intensity of Rayleigh scattering is proportional to λ^{-4}. Rayleigh scattering at atmospheric molecules is responsible for the blue colour of the clear sky.

Refraction Refraction can take place when a beam or pulse travels through a medium with changing properties (i.e. changing refractive index) or when it passes the edge of a solid structure ("side lobes"). The refractive index of air for electromagnetic as well as acoustic waves changes with temperature and humidity. Gradual changes in refractive index (e.g. due to vertical temperature gradients) can lead to considerable bending of acoustic and electromagnetic beams in the atmosphere. Sudden refractive index changes (e.g. at inversions or between different turbulence elements) may lead to backscattering of part of the signal. See also "scintillation"

Remote sensing Most general, remote sensing is measuring from a distance. It relies on receiving incoming radiation, be it acoustic, optical, or electromagnetic signals. Distinction is made between active and passive remote sensing. Passive remote sensing does not influence the state of the measured object. Active remote sensing in some cases provokes specific changes of the measured object (see, e.g. Raman scattering). The opposite technique to remote sensing is in situ measurement.

Scanning Scanning is remote sensing with turnable emitters and receivers. Several scanning techniques can be distinguished. If the azimuth of the instrument is fixed and only the elevation is changed, cross section in vertically oriented planes are obtained. In RADAR meteorology this mode is called range-height-indicator (RHI). If the elevation is fixed and only the azimuth is changed periodically, spatially resolved information from the circumference of the instrument are obtained. In RADAR meteorology this mode is used with low elevation angles and is called plane-position-indicator (PPI). In wind LIDAR applications this mode is used with larger elevation angles and is called conical scanning or VAD (velocity-azimuth-display). Conical scanning is an alternative method to the Doppler beam swinging method. Volume scanning requires the periodical change of both elevation and azimuth.

Scattering When propagating radiation interacts with physical objects (this includes gas molecules, aerosols, and liquid and solid water particles in the atmosphere), or when the refractive index is changing along the path, part of the radiation is scattered. Distinction is made between single scattering (the radiation interacts with only one object) and multiple scattering (several scattering processes take place). See "Mie scattering", "Raman scattering", "Rayleigh scattering" and "Stokes scattering" for further details. Scattering at an angle of 180° is called backscattering.

Scintillation Refractive index fluctuations influencing the propagation of optical or electromagnetic waves along the path from the emitter to the receiver lead to fluctuating intensities recorded by the receiver. This phenomenon, which is responsible for the twinkling of stars at night as well, is called scintillation.

Side lobes If the wavelength of an emitted beam or pulse is comparable or larger than the aperture of the instrument, then refraction takes place at the fringe of the opening and side lobes are produced. Side lobes unavoidably occur with acoustic and electromagnetic remote-sensing techniques, but not with optical techniques. Backscattered signals from side lobes can reach the receiver again and produce fixed echoes (SODAR) or ground clutter (RADAR).

Signal-to-noise ratio The signal-to-noise ratio (frequently abbreviated as SNR) is the ratio between the intensity of the received signal containing wanted physical information and background noise.

Sounding Sounding is an active remote-sensing technique that delivers a range-resolved information. Usually, pulses are emitted and the travel time of the pulses is used to determine the range. The range resolution is determined by the signal propagation speed and the pulse length. Pulses must have some length in order to guarantee meaningful signal-to-noise ratios. The temporal resolution of this remote-sensing method is limited by the travel time from the emitter via the scattering object to the receiver. The opposite measurement technique is path-averaging.

Stokes scattering see "Raman scattering".

Time delay See "travel time".

Time-of-flight See "travel time".

Tomography Tomography is a (usually active) remote-sensing and analysis technique which is able to infer a spatially resolved information from several intersecting path-integrating measurements. It is often called computed tomography (CT) as well. The spatial resolution increases with the number of the beams and the density of their intersection points. Information can only be obtained from the plane which is spanned by the beams.

Travel time A method to determine the distance between an active optical remote-sensing instrument and the backscattering air volume. The time delay or time-of-flight between the emission of a pulse and the receipt of the backscattered pulsed is measured. With the known propagation speed of the pulse, the travel time can be converted into distance information. Alternative range determination methods for continuous wave remote-sensing techniques are beam focusing and frequency modulation.

Virtual temperature The temperature, T_v, which, when entered into the definition for the density of dry air, gives the true density of humid air, is called virtual temperature. This temperature can be computed from the dry-bulb temperature T and the specific humidity q by $T_v = T + 0.609q$. This temperature has been introduced by meteorologists in order to avoid humidity-dependent changes to the gas constant of air. When using the virtual temperature, the gas constant for dry air must be used to compute air density. It cannot be measured directly. See also acoustic temperature.

Voigt line shape The Voigt line shape is a result from a convolution of a Gaussian with a Lorentzian distribution. Doppler broadening leads to a Gaussian line shape, pressure broadening to a Lorentzian line shape.

Wavelet Wavelets are localized orthogonal oscillating functions of limited extend (in contrast to unlimited (nonlocal) trigonometric functions like sine or cosine). A wavelet transform is the representation of a function by wavelets (see, e.g. Kaiser 1996). The word wavelet originates from the French word "ondelette", which means small wave.

References

Bragg WH, Bragg WL (1913) The reflection of x-rays by crystals. Proc Roy Soc Lond A 88: 428–438

Kaiser G (1996) Physical wavelets and RADAR – a variational approach to remote sensing. IEEE Anten Propag Mag 1–20

Schotanus P, Nieuwstadt FTM, de Bruin HAR (1983) Temperature measurement with a sonic anemometer and its application to heat and moisture fluxes. Bound-Lay Meteorol 26:81–93

She C-J (2001) Spectral structure of laser light scattering revisited: bandwidths of nonresonant scattering lidars. Appl Opt 40:4875–4884

Wandinger U (2005) Introduction to LIDAR. In: Weitkamp C (ed.) Lidar. Range-Resolved Optical Remote Sensing of the Atmosphere. Springer Science+ Business Media Inc., New York, 2–18

Name Index

Subject Index

MIX
Papier aus verantwortungsvollen Quellen
Paper from responsible sources
FSC® C105338

If you have any concerns about our products,
you can contact us on
ProductSafety@springernature.com

In case Publisher is established outside the EU,
the EU authorized representative is:
Springer Nature Customer Service Center GmbH
Europaplatz 3, 69115 Heidelberg, Germany

Printed by Libri Plureos GmbH
in Hamburg, Germany